CLIMATE CHANGE AND ITS CAUSES, EFFECTS AND PREDICTION

CARBON LEAKAGE: TRADE ISSUES IN EMISSION REDUCTION

CLIMATE CHANGE AND ITS CAUSES, EFFECTS AND PREDICTION

Additional books in this series can be found on Nova's website under the Series tab.

Additional E-books in this series can be found on Nova's website under the E-book tab.

CARBON LEAKAGE: TRADE IN EMISSION REDUCTION

VICTORIA T. GREENWALD
EDITOR

Nova Science Publishers, Inc.
New York

For permission to use material from this book please contact us:
Telephone 631-231-7269; Fax 631-231-8175
Web Site: http://www.novapublishers.com

Additional color graphics may be available in the e-book version of this book.

LIBRARY OF CONGRESS CATALOGING-IN-PUBLICATION DATA

Carbon leakage : trade issues in emission reduction / editor, Victoria T. Greenwald.
p. cm.
Includes index.
ISBN 978-1-61728-919-4 (hbk.)
1. Carbon dioxide mitigation--Government policy--United States. 2. Carbon dioxide mitigation--Economic aspects--United States. 3. Climatic changes--Government policy--United States. 4. Environmental policy--Economic aspects--United States. 5. United States--Commercial policy. I. Greenwald, Victoria T.
HC110.A4C37 2010
363.738'7460973--dc22

2010027288

Published by Nova Science Publishers, Inc. + New York

CONTENTS

PREFACE

As the debate on reducing greenhouse gases (GHGs) has progressed, increasing concern has been raised about how a U.S. reduction program would interact with programs in other countries. In a global context where currently some countries have legally binding policies to reduce greenhouse gas emission and other countries do not, the potential exists that countries imposing carbon control policies will find themselves at a competitive disadvantage vis-a-vis countries without comparable policies. The risks accompanying establishment of carbon control policies, in the absence of similar policies among competing nations, have been central to debates on whether the U.S. should enact greenhouse gas legislation. This book examines international climate change policy, the potential environmental and economic effects and an analysis of issues raised by the proposed mitigating approaches and options.

Chapter 1- As the debate on reducing greenhouse gases (GHGs) has progressed, increasing concern has been raised about how a U.S. reduction program would interact with those of other countries. In a global context where currently some countries have legally binding policies to reduce greenhouse gas emission and other countries do not—i.e., differentiated global carbon policies—the potential exists that countries imposing carbon control policies will find themselves at a competitive disadvantage vis-à-vis countries without comparable policies.

The risks accompanying establishment of carbon control policies, in the absence of similar policies among competing nations, have been central to debates on whether the United States should enact greenhouse gas legislation. Specifically, concerns have been raised that if the United States adopts a carbon control policy, industries that must control their emissions or that find their feedstock or energy bills rising because of costs passed-through by suppliers may be less competitive and may lose global market share (and jobs) to competitors in countries lacking comparable carbon policies. In addition, this potential shift in production could result in some of the U.S. carbon reductions being counteracted by increased production in less regulated countries (commonly known as "carbon leakage").

Chapter 2 - The United States has proposed, and the European Union (EU) developed, policies to mitigate the potential economic and environmental (i.e., "carbon leakage") impacts of carbon policies on energy- or greenhouse gas-intensive, trade-exposed industries. While studies have found little effect of carbon policies on EU competitiveness in the present, the EU decision to move toward auctioning of allowances in the future has spurred development of criteria to extend potential availability of free allowances to exposed industries to 2020. In a December 2009 decision, the European Commission (EC) listed 164 industrial sectors and subsectors deemed exposed sectors under appropriate European Parliament and Council directives.

Chapter 3 - Changes in the earth's climate attributable to increased concentrations of greenhouse gases may have significant environmental and economic impacts in the United States and internationally. To mitigate climate change effects, countries are taking or considering varying approaches to reducing greenhouse gas emissions, such as carbon dioxide, which is the most important greenhouse gas due to its significant volume. These approaches range from measures to increase energy efficiency to longer-term efforts to develop technologies to establish a less carbon-intensive energy infrastructure. A U.S. policy to mitigate climate change may require domestic production facilities to achieve specified reductions or employ a market-based mechanism, such as establishing a price on emissions.

Chapter 4 – This is a statement of John J. McMackin on behalf of The Energy-Intensive Manufacturers' Working Group on Greenhouse Gas Regulation, before the House Committee on Ways and Means Subcommittee on Trade Hearing on Trade Aspects of Climate Change Legislation.

Chapter 5 - This is a statement of Leo W. Gerard, International President, United Steelworkers, hearing on Trade Aspects of Climate Change Legislation, house Committee on Ways and Means, Subcommittee on Trade.

Chapter 6 – This is a statement of David Hamilton, Director of Global Warming and Energy Programs, Sierra Club, before the Subcommittee on trade of the Committee on ways and Means.

Chapter 7 – This is a statement of Joost Pauwelyn, Professor of International Law, Co-Director of the Centre for Trade and Economic Integration, Graduate Institute of International and Developmental Studies, Geneva (Switzerland) and Senior Advisor, King and Spalding LLP, before the Subcommittee on trade of the House Committee on Ways and Means.

Chapter 8 – This is a statement of Robert E. Clay, CEO and Chairman, Board of Directors of Pridgeon and Clay, Inc., hearing on Trade Aspects of Climate Change Legislation, House Committee on Ways and Means, Subcommittee on Trade.

In: Carbon Leakage: Trade Issues in Emission Reduction ISBN: 978-1-61728-919-4
Editor: Victoria T. Greenwald

Chapter 1

"CARBON LEAKAGE" AND TRADE: ISSUES AND APPROACHES

Larry Parker and John Blodgett

SUMMARY

As the debate on reducing greenhouse gases (GHGs) has progressed, increasing concern has been raised about how a U.S. reduction program would interact with those of other countries. In a global context where currently some countries have legally binding policies to reduce greenhouse gas emission and other countries do not—i.e., differentiated global carbon policies—the potential exists that countries imposing carbon control policies will find themselves at a competitive disadvantage vis-à-vis countries without comparable policies.

The risks accompanying establishment of carbon control policies, in the absence of similar policies among competing nations, have been central to debates on whether the United States should enact greenhouse gas legislation. Specifically, concerns have been raised that if the United States adopts a carbon control policy, industries that must control their emissions or that find their feedstock or energy bills rising because of costs passed-through by suppliers may be less competitive and may lose global market share (and jobs) to competitors in countries lacking comparable carbon policies. In addition, this potential shift in production could result in some of the U.S. carbon reductions being counteracted by increased production in less regulated countries (commonly known as "carbon leakage").

There are three basic approaches, which are not mutually exclusive, to assist greenhouse gas- intensive, trade-exposed industries: (1) directly supporting domestic industries; (2) penalizing foreign competitors; and (3) developing alternative sectoral approaches. Importantly, these are presumably transitional actions, pending some international agreement that "levels the playing field."

Each approach has its own focus. Support for domestic industries, embodied in most legislative proposals, is focused on preserving the industry's current competitive position and jobs and may, depending on the details, help transition that industry to the future. It does not directly promote an international agreement. Trade measures levied against foreign competitors, another approach being proposed, may provide a stick for international negotiation, but the primary focus is on protecting greenhouse gas-intensive, trade-exposed industries from "unfair" competition— producers in countries not imposing comparable carbon control policies. Finally, the sectoral approach represents a range of options focused on integrating developing countries' industrial base into a mutually acceptable international framework that provides a level playing field for all participants. Whether any of these approaches would have any appreciable effect on carbon leakage is unclear.

The design of an assistance program—the goals, eligible participants, implementation and enforcement—would be difficult to define in a manner that satisfies all parties. There is every incentive for any industry facing a cost increase from carbon policies to claim that its competitive position could be diminished, thereby justifying special consideration by the government. The government would be in the difficult position of picking winners and losers, sometimes without access to important, but proprietary, data.

INTRODUCTION

As the debate on reducing greenhouse gases (GHGs) has progressed, increasing concern has been raised about how a U.S. reduction program would interact with programs in other countries. In a global context where currently some countries have legally binding policies to reduce greenhouse gas emission and other countries do not—i.e., differentiated global carbon policies—the potential exists that countries imposing carbon control policies will find themselves at a competitive disadvantage vis-à-vis countries without comparable policies.

The risks accompanying establishment of carbon control policies, in the absence of similar policies among competing nations, have been central to debates on whether the United States should enact greenhouse gas legislation. Specifically, concerns have been raised that if the United States adopts a carbon control policy, industries that must control their emissions or that find their feedstock or energy bills rising because of costs passed-through by suppliers may be less competitive and may lose global market share (and jobs) to competitors in countries lacking comparable carbon policies. In addition, this potential shift in production could result in some of the U.S. carbon reductions being diluted by increased production in more carbon intensive countries (commonly known as "carbon leakage").[1]

In response to these concerns, several proposals introduced in Congress would attempt to mitigate the effect of carbon policies on affected U.S. industry. Proposed mitigating actions include, for example, providing assistance to greenhouse gas-intensive, trade-exposed industries, or imposing tariffs on certain greenhouse gas-intensive goods imported into the country from countries not implementing comparable carbon policies.

This chapter examines the dynamics of this issue in three parts: (1) exploration of the nature of the problem with respect to international climate change policy, potential environmental effects, and potential economic effects; (2) identification of a range of possible options to address concerns; (3) analysis of issues raised by the proposed mitigating approaches and options; and (4) implications of the various approaches.

Nature of the Problem

There are three components of the problem of differentiated global carbon policies with respect to trade: (1) the lack of an international agreement with binding targets to reduce greenhouse gases; (2) the issue of carbon leakage; and (3) economic and competitive effects. Each of these is discussed below.

Lack of Global Agreement to Regulate Greenhouse Gases

For those policymakers who argue that human activities have changed or threatened to change the global climate, the policy debate on a U.S. climate change strategy has revolved about three major considerations: the posited reduction scheme's *cost* of compliance, its impact on the country's *competitiveness*, and its *comprehensiveness* with respect to developing countries who currently have no binding reduction targets.[2] These three considerations (the "three Cs") are interlinked, especially the international aspects of competitiveness and comprehensiveness. That no international agreement addresses the international competitiveness and comprehensiveness issues has led to a major debate in the Congress about whether to include unilateral trade provisions, targeted subsidies, or other provisions in any domestic greenhouse gas reduction scheme to address them.

It should be emphasized that this debate results from the lack of a comprehensive, international agreement to mandate strategies to reduce greenhouse gas emissions. The most effective and efficient solution, both economically and environmentally, would be a comprehensive agreement. Climate change is a global problem ultimately requiring a global solution. Any unilateral solution considered necessary would probably be temporary and transitional in nature. As stated by the Australian Government in its green paper on reducing carbon emissions:

> The first best solution to address the competitive concerns of EITE [emissions-intensive trade-exposed] industries[3] would be to develop a comprehensive global agreement under which all major emitters have binding carbon constraints. Effective sectoral agreements for EITE industries would also address these concerns for industries covered by such agreements. However, in the absence of these developments, assisting EITE industries in response to the introduction of the scheme may be warranted on environmental grounds and because it may smooth the transition of the economy.[4]

However, it is not clear when such an agreement will be concluded and whether it would be acceptable to the United States. International working groups set up under the Bali "Action Plan" to develop a "Post-Kyoto" agreement are scheduled to present their results at the Copenhagen meeting of the Conference of the Parties to the United Nations Framework Convention on Climate Change (UNFCCC) (COP-15) and the Meeting of the Parties to the Kyoto Protocol scheduled for November 30 -December 11, 2009. A successful conclusion to these ongoing efforts leading up to that conference could render this issue moot.

Environmental Issue: Carbon Leakage

Although carbon leakage is generally defined in terms of differentiated carbon policies and their resulting impacts on greenhouse gas emissions, the phenomenon is much more complicated, involving differences in countries' economies (such as labor costs and exchanges rates) and trade flows among them.[5] Thus carbon leakage, like the job leakage issues discussed later, is an interaction that will continue regardless of whether carbon policies are enacted. The focus here is on minimizing carbon leakage resulting specifically from differentiated carbon policies.

In the context of analyzing the effect of differentiated carbon policies, carbon leakage is a twofold problem. The first is the possibility that introduction of a carbon control regime in a country ahead of the introduction of a comparable policy in competing countries could result in the production of greenhouse gas-intensive products diminishing in the country attempting to control emissions and increasing in competing countries with no carbon controls. Basically, countries with carbon controls risk losing global market share to competing countries without controls. This would counteract the net reductions achieved by the country attempting to address climate change and reward economically the countries that were not.

The second problem is a longer-term possibility that future investments by greenhouse gas- intensive industries could be channeled to countries with no (or less stringent) carbon controls, circumventing carbon reduction needs and potentially locking in obsolete technology. This relocation and construction of new facilities without carbon control could make future reductions more difficult and expensive.

Studies of potential carbon leakage resulting from strategies to reduce greenhouse gases have produced a range of estimates. The only attempt to estimate the leakage impact of proposed U.S. legislation is the Environmental Protection Agency's (EPA) analysis of the Lieberman-Warner Climate Security Act of 2008 (S. 2191) in the 110th Congress.[6] Not surprisingly, a leakage estimate for the year 2050 is highly dependent on assumptions about the U.S. economy, international actions to reduce emissions, and U.S. reduction strategies.[7] EPA found that if non- Annex I countries[8] were to adopt a greenhouse gas reduction target beginning in 2025 that holds their emissions at 2015 levels through 2034 and then further reduce their emissions to 2000 levels thereafter, then no emission leakage would occur under the proposed legislation. This result emphasizes the above point that the most effective solution to the leakage problem would be a long-term agreement to incorporate developing countries into an international accord on greenhouse gas emissions.

EPA also conducted a sensitivity analysis assuming no greenhouse gas reductions by non-Annex 1 countries through 2050. Under this scenario, leakage of U.S. reductions were estimated at about 11% in 2030, and 8%-9% in 2050.[9] EPA notes that part of the reason for the somewhat modest leakage rates estimated by the model is the significant demand by Annex I countries for international credits from non-Annex I countries, reducing their emissions.

Leakage has also been studied by the European Union (EU) with the implementation of its Emissions Trading Scheme (ETS).[10] In general, there has been little indication of any leakage resulting from phase 1 of the ETS. A variety of explanations are possible, including strong demand for aluminum and other commodities that has allowed manufacturers to pass on costs and remain profitable, the short time-frame (2005-2007) that makes it difficult to discern potential long-term investment trends, and the efforts of individual EU members to protect their industries through free allowance allocations.

Studies suggest that leakage may be a longer-term issue as more stringent reduction targets are imposed. Analysis of the EU's climate change package that would lead to a 20% reduction in greenhouse gases from 1990 levels by 2020 (referred to as the post-2012 program) has produced a range of carbon leakage estimates. As indicated by Table 1, assumptions about technology development and spillover effects, elasticity of energy supply, and the mobility (substitution) of energy-intensive production between countries produces estimates that make conclusions about the carbon leakage effects of a very aggressive reduction target difficult to assess.

The Intergovernmental Panel on Climate Change (IPCC) has also weighed in on the carbon leakage debate with respect to the short-term Kyoto Protocol commitment period (2008-20 12). In its 2007 assessment, the IPCC makes three observations with respect to carbon leakage:

- Model-based estimates of "carbon leakage" from implementing Kyoto Protocol commitments are in the range of 5%-20% (i.e., 5%-20% of domestic reductions may be offset by displacement abroad) *(IPCC confidence in conclusion: medium agreement, medium evidence)*
- Empirical studies on energy-intensive industries under the EU-ETS conclude that carbon leakage is "unlikely to be substantial" due to transport costs, local market conditions, product specialization of local suppliers, etc. *(IPCC confidence in conclusion: medium agreement, medium evidence)*

Table 1. Estimated Potential Carbon Leakage from Implementing the EU's Proposed Post-2012 Program

Study	Carbon Leakage	Remarks
C. Di Maria and E. van der Werf (2005)	Unconstrained cou-ntry will voluntarily decrease emissions	Results depend on directed technical change
R. Gelagh and O. Kuik (2007)	15% to -15%	Rates depend on development of technology and spill-over
J.M. Burniaux, and Oli-veira Martins (2000) and J.M. Burniaux, R. McD-ougall, and T.P. Truong (2002)	2% to 5%	Assumes inelastic supply of energy: less energy demand results in lower prices
M.H. Babiker (2001), (2005), with H.D. Jacoby (1999)	>100%	Assumes perfect substitution of energy-intensive industrial production between different countries
J. Bollen, T. Manders, and H. Timmer	14%	Assumes inelastic supply of energy: less energy demand results in lower prices

Source: Based on Christian Lutz and Ulrike Lehr, "Employment Effects within the Climate Change Policy Framework," in European Parliament, Policy Department, *Competitive Distortions and Leakage in a World of Different Carbon Prices* (July 2008), Table 5.2.

- Quantifying possible benefits of international transfer of low carbon technologies induced by industrialized country action is not possible.[11]

Economic Issue: International Competitiveness

Competitiveness can be a rather abstract term for which any precise meaning can be elusive. As with carbon leakage, competitiveness is a continuing phenomenon, with companies becoming more or less competitive according to a host of factors, including productivity, market demand, resource costs, labor costs, exchange rates, and the like. As stated by the Australian Government in its Green Paper on carbon reduction schemes:

> Changes in the cost structures of entities and industries are not unusual and occur continuously in a market-based economy; nor is it unusual for Government policy to change cost structures. For example, the adoption of high quality occupational health and safety standards have affected the profitability of Australia's labour-intensive traded industries, making it more difficult for them to compete with foreign producers that are subject to lower standards. Assistance is not usually provided to offset the impact of domestic policies on traded industries, as those policies reflect the priorities and values of the Government and community more generally.[12]

Most industries face a competitive market (sometimes international in scope) both in terms of producers of the same products and producers of substitute products. Also, in some cases, an industry may face a fairly elastic demand for its product. Thus, most industries are price sensitive, and therefore any increase in manufacturing costs – as by a carbon emission reduction requirement – hurts the competitiveness of a firm. This complex situation is further complicated for energy-intensive industries as competitors within the same industry may experience different energy price increases (particularly for electric power), depending on their individual energy needs and power arrangements. For example, an aluminum plant receiving power from a hydroelectric facility may not be affected the same way as a similar plant whose power contract is with a coal-fired power supplier.

Such differences among individual companies could have several potential impacts. First, as noted above, it may affect the competitive balance of specific domestic facilities. Second, investment decisions by industries could be affected, particularly with respect to technology. New, more efficient technology is emerging for some processes. The combination of high, but volatile, price signals being sent from the energy markets and potential ones from a carbon policy could speed their development. If commercialized, new technology could reduce the impact of any carbon policy and, indeed, could improve competitiveness. Analysis in sufficient sector- specific detail to examine this possibility, or to develop proxies to explore the possibilities for industry technology over the next 40 years, are beyond the scope of this chapter.

A company's ability to compete under a carbon policy depends on three primary factors: (1) the greenhouse gas intensity of a company's products which influences the company's profitability and the products' cost; (2) the company's ability to pass on any increased costs to consumers without losing market share or profitability; and (3) the company's ability to mitigate carbon emissions, reducing the impact of the carbon policy on its operations and profitability.[13] Each of these factors involves a web of site-specific interactions.

Greenhouse-Gas Intensity

An industry's greenhouse-gas intensity factor is a foundation both of any direct greenhouse gas emissions produced by the manufacturing process of the product (e.g., PFCs from aluminum production, CO2 from cement manufacture), and of any indirect greenhouse gas emissions produced by the inputs to the manufacturing process (e.g., electricity, natural gas). Much of the discussion of greenhouse gas-intensive industries is in fact a discussion of energy-intensive industries. However, as noted above, this is an imperfect indicator as different plants will have different energy sources and, thus, different indirect greenhouse gas emissions. In addition, such a focus ignores the 320 million metric tons of annual greenhouse gas emissions that U.S. industrial processes emit directly.

That the impact of a carbon policy on product prices, employment, and profitability is dependent on its greenhouse gas intensity is seemingly straightforward. However, the measurement of such intensity may not be. Metrics that could be used to determine carbon intensity include employment per unit of emissions, value added by the production activities per unit of emissions, or revenue generated by the activity per unit of emissions.[14] Each indicator differs in level of transparency, variability over time and within sectors, and emphasis on scheduling of capital structure and labor needs. Choosing an indicator or combination of indicators that all parties believe fairly represent the industries of concern would be challenging.

As suggested above, industries can be greenhouse gas-intensive from either the process they employ (direct emissions) or the energy fed into the process from outside (indirect emissions), or both. The greenhouse gas intensity can be measured in terms of its impact on product price, company profitability, or labor. Most studies of greenhouse gas-intensive industries actually focus on energy-intensive industries.

As suggested by Table 2, the complexity of determining carbon intensiveness is significantly influenced by the level of sector aggregation one chooses to focus on. For example, while several 3-digit North American Industrial Classification System (NAICS) industry categories, like paper, chemicals, and nonmetallic mineral products, have aggregated energy costs of less than 8% of value, 6-digit NAICS industry subcategories, such as pulp mills, newsprint mills, alkalies and chlorine, nitrogenous fertilizers, lime, and primary aluminum shelters, have energy costs approaching or exceeding 20% of value.

In addition, a single product may exhibit highly variable emissions, depending on the technology used. For example, Figure 1 provides International Energy Agency (IEA) data on average carbon dioxide emission per ton of crude (or raw) steel manufactured by several different processes or energy sources. As indicated by the blue bars, the process used to manufacture steel has a substantial effect on the direct and indirect CO_2 emissions emitted. In addition, emissions are influenced by the processes' source and consumption of electricity. As indicated by the red arrows, indirect emissions from electricity sources have a significant effect on the total emissions from a given process.

Table 2 provides data on the energy-intensiveness of an illustrative set of manufacturing industries. Two metrics are displayed. The first measures the importance of energy costs to the total value of the industry's products. The second measures the importance of energy costs per person employed by the industry.

Table 2. 2002 Data on Energy-Intensiveness of Manufacturing Sector

Industry (NAICS code)	Energy costs as share of value	Energy cost per employee
Food and beverage (311, 3121)	1.49%	$5,324
Textiles (313, 314)	2.40%	$4,747
Apparel (315)	1.01%	$1,202
Wood products (321)	1.66%	$2,930
Paper (322)	7.27%	$24,082
Pulp mills (322110)	*21.73%*	*$95,881*
Paper mills, except newsprint (322121)	*9.74%*	*$45,037*
Newsprint mills (322122)	*18.89%*	*$90,430*
Paperboard mills (322130)	*17.30%*	*$76,458*
Printing (323)	1.38%	$1,914
Petroleum refineries (324110)	7.39%	$231,865
Chemicals (325)	4.28%	$24,268
Petrochemicals (325110)	*12.39%*	*$268,881*
Alkalies and chlorine (325181)	*31.79%*	*$146,205*
Carbon black (325182	*15.50%*	*$84,495*
Other basic inorganic chemicals (325188)	*6.87%*	*$24,396*
Basic organic chemicals (325199)	*11.47%*	*$67,194*
Plastic materials and resins (325211)	*7.16%*	*$43,962*
Nitrogenous fertilizers (325311)	*19.19%*	*$152,334*
Pharmaceuticals and medicines (3254)	*0.66%*	*$4,356*
Nonmetallic mineral products (327)	5.45%	$11,347
Glass (3272)	*6.06%*	*$12,255*
Cement (327310)	*16.58%*	*$71,296*
Lime (327410)	*23.23%*	*$57,016*
Ferrous metals (331111, 331112, 3312, 331511)	8.81%	$30,039
Iron and steel mills (331511)	*11.62%*	*$47,207*
Iron foundries (331511)	*6.44%*	*$10,237*
Nonferrous metals (3313, 3314, 331521, 331524)	4.79%	$13,570
Primary aluminum smelters (331312)	*19.83%*	*$83,222*
Aluminum foundries (331524)	*3.51%*	*$6,074*
Other nonferrous metals (3314)	*2.87%*	*$9,598*
Fabricated metal products (322)	1.77%	$2,685
Machinery (333)	0.80%	$1,792
Computers and electronics (334)	0.46%	$1,304
Electrical equipment (335)	0.68%	$1,445
Transportation equipment (336)	0.60%	$2,396
Furniture and related products (337)	0.79%	$1,003

Source: Compiled by Peterson Institute and World Resources Institute from: U.S. Department of Energy, Energy Information Administration, *Manufacturing Energy Consumption Survey* (2002).

Raw steel production is also the most CO2-intensive step in steel production. Figure 2 provides illustrative data from the United Kingdom (UK) on the value chain of an integrated steel production process through its various steps using a Basic Oxygen Furnace (BOF) to make its raw steel. In a BOF, iron ore is reduced to semi-finished steel, which is subsequently hot rolled and then further refined into specific finished products. Semi-finished steel production is the most carbon-intensive and electricity-intensive step in integrated raw steel production. In contrast, the value produced at this step is relatively low compared with the emissions. This ratio (called the product value at stake (VAS)) means that the step would incur high CO_2 cost increases relative to product value. The extent to which these costs are spread across the subsequent production steps would lower the overall cost impact on final production.[15]

This suggests that the primary competitiveness issue with a BOF is with the semi-finish step of raw steel production. This is illustrated in the figure by comparing the difference between the total cost increase (solid line) and the semi-finished steel increase (dashed line). The relatively small increment of increase created by the downstream processes compared with semi-finished steel production suggests the dominant effect of CO_2-induced cost increases from semi-finished steel production on downstream production cost increases.[16]

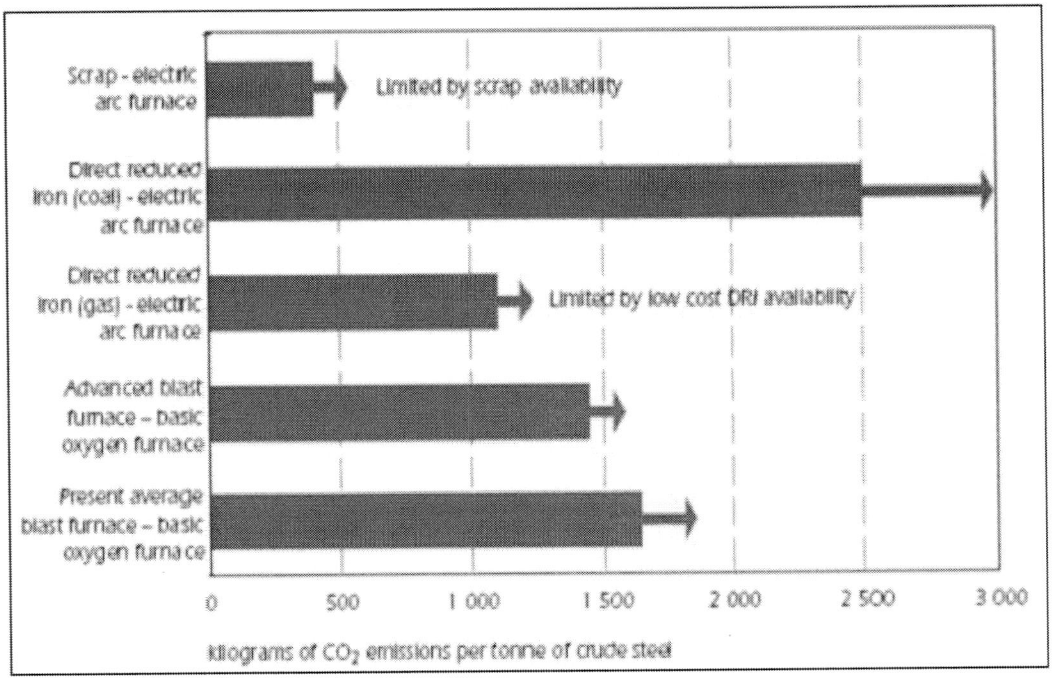

Source: International Energy Agency, *Tracking Industrial Energy Efficiency and CO_2 Emissions (2007),* p. 108.

Note: The high and low-end ranges indicate CO_2-free and coal-based electricity, and account for country average differences based n IEA statistics. The range is even wider for plant based data. The product is crude steel, which excludes rolling and finishing.

Figure 1. Carbon Dioxide Emissions Per Ton of Crude Steel

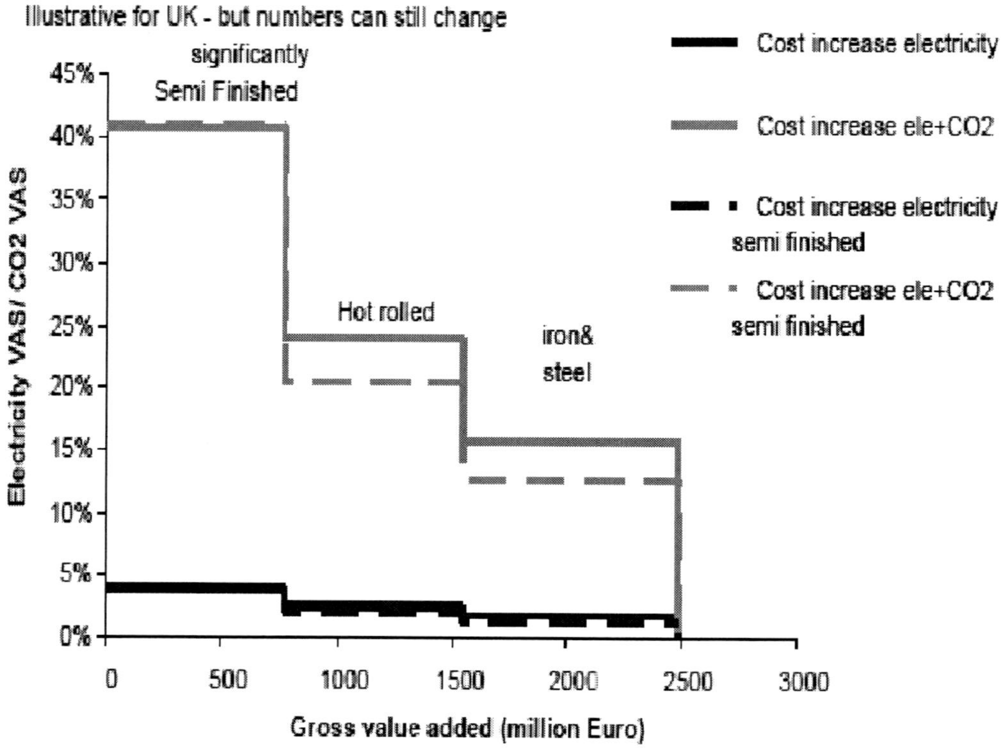

Source: Kasten Neuhoff and Susanne Droege, *International Strategies to Address Competitiveness Concerns* (July 2007), p. 4.

Figure 2. Value Chain of Steel Production

What Table 2, Figure 1, and Figure 2 do not indicate is the substantial difference in emissions due to site-specific considerations, such as age of plant, maintenance, etc., that would make the ranges greater than presented here. The figures also do not indicate the varying degrees of product integration that steel mills may include. For example, most BOFs combine crude steel making with hot rolling to avoid additional energy consumption in repeated heating cycles. The more a plant combines production steps, the less the overall cost effect of a carbon policy.[17] Also, the link between the BOF and hot rolling plant may make relocation more difficult although, if it occurs, the affected community would lose both the BOF and the hot rolling plant.

Also, as suggested by Figure 1 the processes are not completely substitutable, even if the crude steel is. For example, the use of electric arc furnaces is constrained by the availability of scrap steel, particularly in developing countries such as China.

Cost Pass-through Ability

A sector's ability to pass through the cost of carbon policies is similarly differentiated. The ability of companies to pass through costs from carbon policies primarily depends on three factors: (1) the price-responsiveness of demand for the product; (2) market structure and dynamics that include the number of competitors and amount of regulation and state-ownership; and (3) the international scope of the competition, particularly with respect to differentiated carbon policies.[18] For example, the electricity sector can generally pass on its costs to consumers because electricity demand is relatively price-inelastic, the market structure is significantly regulated, and there is very limited international competition from countries with no carbon policies. Chlorine, produced with a very electricity-intensive process, is a hazardous substance that could raise serious transport issues, potentially reducing the ability to substitute foreign production for domestic production. In contrast, other sectors, such as raw steel, are in very competitive markets with significant international trade exposure (although during periods of high demand for steel and other primary metals, prices have risen substantially).

Elasticity of Demand

Elasticity of demand refers to how people respond to an increase in a product's price. Inelastic price behavior by consumers indicates that companies can raise prices in response to increased costs without a substantial response by consumers to reduce use of the product or seek a substitute. Elastic price behavior by consumers means they are sensitive to price increases and will seek to either reduce demand for a product or seek a substitute. Companies facing an elastic demand for their products because of available substitutes would have a more difficult time passing on any cost increases resulting from carbon policies. In contrast, companies facing an inelastic demand for their products would have more flexibility in addressing the same cost increases.

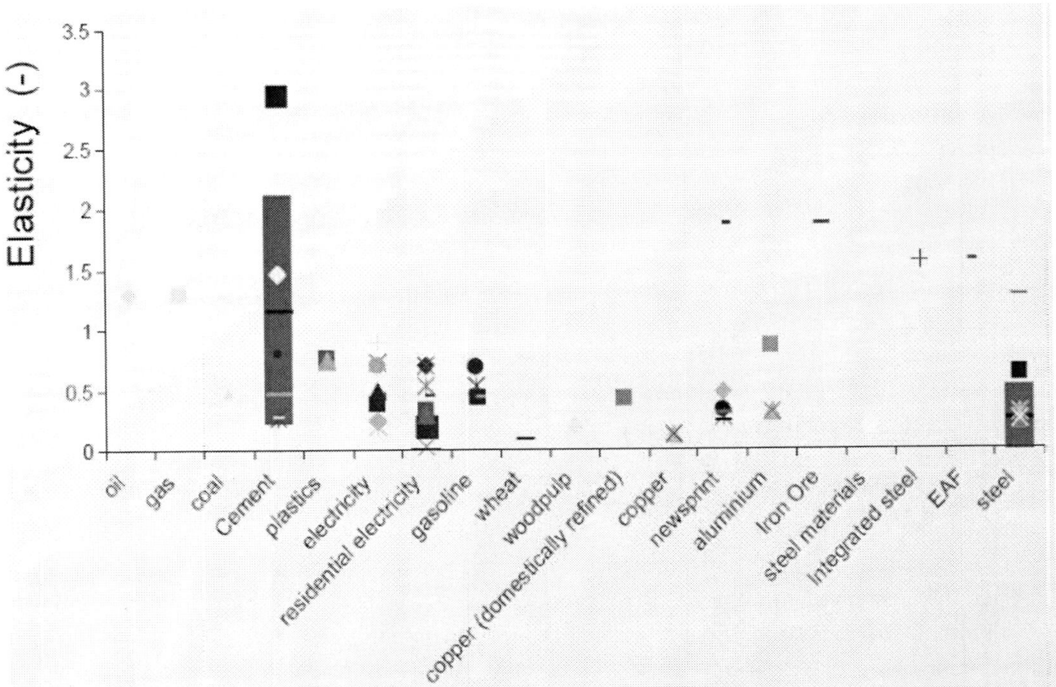

Source: Misato Sato and Karsten Neuhoff, *Testing the Demand Substitution Effect* (June 22, 2007), p. 5. Symbols represent the various studies and estimates included. See source for details.

Figure 3. Literature Survey of Price Elasticities of Demand

Based on a review of the literature and their own estimating methodology, Sato and Neuhoff estimated the short and long run effects of price changes on demand for various commodities in the European context.[19] These estimates are presented in Figure 3. The authors found that demand for electricity and several other commodities appear to be relatively inelastic (less than -1) while demand for cement and for steel products from some processes may be fairly elastic. However, as indicated, there is considerable uncertainty in these estimates and they should be considered indicative of the importance of pricing to consumption of these commodities and not predictive. Indeed, the range presented suggests that quantifying this variable as part of any eligibility criteria for an assistance program may be difficult.

Market Structure

The number and concentration of firms in a given market and the extent of government involvement and regulation of that market influence the ability of firms to pass on costs. For example, public utilities that are regulated by a public service commission are generally allowed to pass-through any legitimate cost increases to consumers. Likewise, industries where a few companies have concentrated market power to influence prices may have an enhanced ability to pass through costs through their ability to influence prices.

To illustrate the degree of concentration in various parts of the manufacturing sector, Table 3 provides two indicators of market concentration across an illustrative sample of the U.S. manufacturing sector. The first is the market share of the Top 4 companies in a category. The second is the Herfindahl-Hirschman Index (HHI) for the Top 50 companies in the same category. The HHI is a commonly accepted measure of market concentration and is used by the Department of Justice (DOJ) in reviewing mergers and acquisitions for potential anti-trust concerns.[20] An HHI between 1000 and 1800 is considered moderately concentrated by the DOJ, and an HHI in excess of 1800 is considered concentrated. Transactions that increase the HHI by more than 100 points in concentrated markets presumptively raise DOJ anti-trust concerns.

As indicated by Table 3, the broader industrial categories (3-digit NAICS) would suggest that market concentration is not a major issue with respect to pass-through ability. However, as suggested by the previous discussion of greenhouse gas intensity, disaggregating a sector can reveal a more complex situation. For example, the NAICS 3-digit Chemical category suggests little concentration in that sector. However, a sampling of subcategories indicate several that are at least moderately concentrated. There is a similar situation for primary metals. If the categories were disaggregated further to include categories such as glass container manufacturing (327213) or electrometallurgical ferroalloy product manufacturing (331112), more pockets of concentration would be found. Thus, companies can be more able to pass through costs in some of their products than in others. Therefore, unless eligibility requirements for any government assistance are sufficiently detailed to direct aid only to those categories that can not raise prices, the government risks providing support for companies that don't need it. Of course, the potential ability of companies to pass-through cost increases can be muted by international competition, as discussed next.

Table 3. 2002 Data on Market Concentration in various parts of the Manufacturing Sector

Industry (NAICS code)	Market Share: Four Largest	HHI of 50 Largest[a]
Apparel (315)	17.3%	105.7
Wood products (321)	10.0%	48.4
Paper (322)	25.8%	259.3
-Pulp mills (322110)	*61.1%*	*1175.2*
-Paper mills, except newsprint (322121)	*53.1%*	*810.2*
-Newsprint mills (322122)	*53.9%*	*976.6*
-Paperboard mills (322130)	*48.5%*	*748.5*
Printing (323)	10.4%	45.2
Petroleum refineries (324110)	41.2%	639.7
Chemicals (325)	13.7%	99.9
-Petrochemicals (325110)	*84.7%*	*2661.6*
-Alkalies and chlorine (325181)	*73.2%*	*1786.4*
-Carbon black (325182)	*76.0%*	*1791.8*
-Other basic inorganic chemicals (325188)	*20.9%*	*216.9*
-Other basic organic chemicals (325199)	*22.0%*	*238.3*
-Plastic materials and resins (325211)	*32.4%*	*442.5*
-Nitrogenous fertilizers (325311)	*53.9%*	*976.9*
-Pharmaceuticals and medicines (3254)	*34.0%*	*506.0*
Nonmetallic mineral products (327)	7.0%	46.7
-Glass (3272)	*24.5%*	*278.0*
-Cement (327310)	*38.7%*	*568.5*
-Lime (327410)	*66.3%*	*1254.5*
Primary metals (331)	20.0%	149.
-Iron and steel mills (331111)	*44.4%*	*656.7*
-Iron foundries (331511)	*29.2%*	*350.1*
-Primary aluminum smelters (331312)	*85.3%*	*D*
-Aluminum foundries (except die-casting) (331524)	*25.0%*	*267.4*
-Nonferrous metals (except aluminum) (3314)	*21.1%*	*213.4*
Fabricated metal products (332)	3.7%	10.2
Machinery (333)	14.4%	71.3
Computers and electronics (334)	18.0%	135.0
Electrical equipment (335)	16.5%	113.9
Transportation equipment (336)	42.1%	574.7
Furniture and related	11.0%	57.2

Source: U.S. Census Bureau, Economics and Statistics Administration, *Concentration Ratios: 2002* (May 2006).

Note: D = withheld to avoid disclosing data of individual companies.

Herfindahl-Hirschman Index.

Trade Exposure

That an industry has some trade exposure does not necessarily mean that it would be hurt under a carbon reduction policy. The key aspect of trade exposure in terms of carbon policy is whether a sector is considered a price-taker on world markets. If the price of its product is dictated by world supply and demand, then its ability to raise prices may be constrained. This situation could result in the sector deciding to reduce domestic production in the short term and moving factories overseas in the long run.

One measure of trade exposure is the penetration of imports as a share of total U.S. demand for a product because it indicates the availability of foreign substitutes for that product.[21] Table 4 provides the 2006 import share of demand for a variety of activities. Because of the aggregation issue identified above, this is only a rough indicator of sectors that could have difficulty passing on cost increases because of international competition. As indicated, several greenhouse gas- intensive sectors have significant import penetration, including primary metals (nonferrous and ferrous), basic chemicals, and finished products that use these commodities, including electronics, machinery, and transportation. Other greenhouse-intensive sectors, such as cement, lime, and paper have less penetration in the aggregate.

Table 4. 2006 Data on U.S. Market Share of Imports for Various Parts of the Manufacturing Sector

Sector	Import Share of Demand	Average WTO Applied Tariff
Apparel	74.8%	16.6%
Electronics	51.5%	7.1%
Nonferrous Metals (primary aluminum smelters, aluminum foundries, others)	42.8%	7.4%
Machinery	37.2%	5.7%
Transportation	34.4%	7.4%
Textiles	27.3%	9.5%
Furniture	24.9%	9.8%
Ferrous Metals (iron and steel mills, iron foundries)	23.3%	6.1%
Chemicals (petrochemicals, alkalies and chlorine, other organic and inorganic chemicals, etc.)	22.3%	5.0%
Wood Products	17.8%	6.4%
Refining	15.7%	3.9%
Nonmetallic Mineral Products (glass, cement, lime)	14.7%	8.7%
Plastics	14.0%	8.8%
Fabricated Metals	13.9%	9.5%
Paper (pulp and paper mills)	13.6%	3.6%
Food and Beverage	6.8%	17.3%
Printing	5.9%	3.8%

Source: Compiled by Peterson Institute and World Resources Institute from: U.S. Department of Commerce, *Bureau of Economic Analysis, Industry Economic Accounts* (2007), and from the World Trade Organization.

A second measure of trade exposure competition is the tariffs countries have placed on imports to protect their sector from international competition, fair or unfair. In some sense this measure indicates the perceived threat that international competition presents to the viability (and thus, potential relocation) of domestic production. The metric presented in Table 4 is the average tariff rate applied on that sector's products by the 15 largest members of the World Trade Organization (WTO). As indicated, this metric suggests that labor-intensive industries, such as apparel, textiles, and furniture have received the most attention in tariff determinations. Among greenhouse gas- intensive sectors, nonmetallic mineral products (glass, cement, and lime) have received the most attention, followed by nonferrous metals, such as aluminum, while paper received the least attention.

Ability to Respond to Carbon Policy

A company's ability to respond to carbon policies depends on the alternatives available and on the timing and costs of mandated action. Although, as noted earlier, a comprehensive review of carbon policy options for industry is beyond the scope of this report, a recent report by McKinsey & Company illustrates some of the cost and potential for reducing greenhouse gas emissions by industry. The report found that significant cost-effective reductions can be achieved by 2030.[22] A summary of that potential is provided in Figure 4 below.

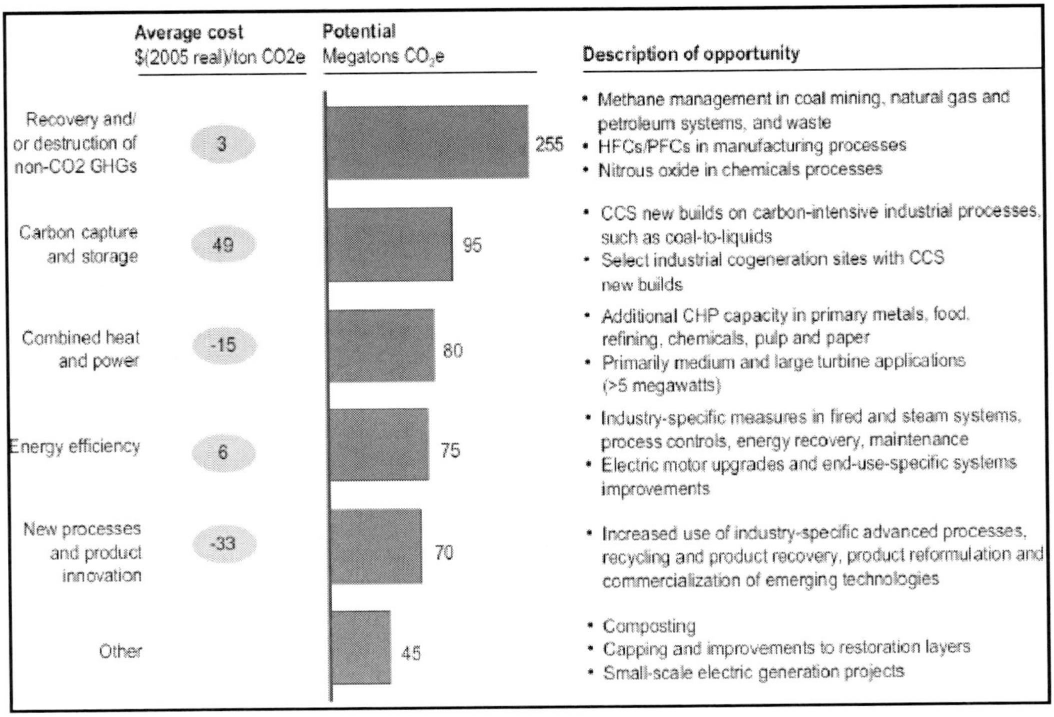

Source: McKinsey & Company, Reducing U.S. Greenhouse Gas Emissions: How Much at What Cost? (December 2007) p. 50.

Figure 4. Abatement Options for U.S. Industrial and Waste Cluster:2030 Mid-Range Case(options under $50/ton CO_2e)

Although the report identifies opportunities for reductions by 2030, it also notes the problem of fragmentation. Specifically, it notes that much of the abatement potential is spread over 75+ options and dependent on either favorable economics or regulatory support. As stated in the report:

> Although the reference case assumes that improvements in the energy intensity or processes in some sub-sectors (e.g., aluminum, food, cement) will avoid some 470 megatons of future emissions, these improvements are not assured and still must be captured. Without supportive regulatory structures, some of these improvements may not be made or the emission will be "off-shored" to other economies, with U.S. domestic GHG emissions decreasing and global emissions staying flat or rising.[23]

Timing also affects the availability of options and the ability of industry to respond to carbon policy. Indeed, all of the factors identified here are affected by the timing of any carbon policy. As stated in a joint Peterson Institute for International Economics and World Resources Institute report:

> In the short term, most firms have limited ability to improve the efficiency of capital stock or switch to alternative sources of energy. How much of the energy cost increase the firm must absorb then depends on the immediate availability of substitutes for the firm's products. Over the medium and long terms, firms have greater ability to seek out lower-carbon fuel sources and develop more energy-efficient technology.[24]

Options to Provide Assistance

There are three basic approaches to assisting greenhouse-gas-intensive, trade-exposed industries: (1) assist domestic industry; (2) penalize foreign competitors; and (3) develop alternative sectoral approaches. It should be noted that these categories are not mutually exclusive; all three could be used, either in combination for a given industry, or separately as appropriate to a given industry's characteristics and needs.

Assist Domestic Industry

In some ways, the simplest approach is to assist domestic industry to compensate for the negative economic effects of carbon policies. Depending on the carbon scheme approach (cap-and-trade, carbon tax, regulations, etc.) the assistance could be in the form of (1) free allocation of allowances (cap-and-trade program), (2) tax credits (carbon tax), or (3) cash payments (any approach). To the extent that many carbon regimes include substantial support for research and development (R&D), this approach is already incorporated in the overall debate. Federal support for R&D also could be considered as one approach to the objective of encouraging a smooth transition to a less-carbon-intensive industry. This discussion explores options that go beyond R&D in efforts to assist greenhouse gas-intensive, trade-exposed industries.

Free Allocation of Allowances under a Cap-and-Trade Regime

A cap-and-trade program is based on two premises. First, a set amount of greenhouse gases emitted by human activities can be assimilated by the ecological system without undue harm. Thus the goal of the program is to put a ceiling, or cap, on the total emissions of greenhouse gases. Second, a market in pollution licenses between polluters is the most cost-effective means of achieving a given reduction. This market in pollution licenses (or allowances, each of which is equal to 1 ton of carbon dioxide equivalent) is designed so that owners of allowances can trade those allowances with other emitters who need them or retain (bank) them for future use or sale. Allowances may be allocated free by the federal government to affected entities or other parties, or auctioned by the government for a variety of purposes.

Creating an allowance system is similar to creating a new currency. The allowance has value that can be converted to cash via a market clearing mechanism, such as an exchange. Thus, allocating allowances for free is essentially the same as distributing money or assistance to affected parties.

Tax Credits under a Carbon Tax Regime

A program similar to a free allocation approach under a cap-and-trade scheme can be achieved under a carbon tax regime. Instead of providing industries with free allowances, a tax credit program would provide tax credits to them. For industries with substantial direct greenhouse gas emissions, the tax credit approach would be very straightforward as both the eligible emissions and the carbon tax would be well known (unlike free allowances where the precise value of the allotment can only be estimated beforehand). For industries with substantial indirect emissions, the process would be more involved as the eligible emissions for credit would have to be estimated; an estimate that would depend on industry, source of energy, and process involved. Like a free allocation system, the system could be phased out over time in order to encourage a smooth transition to a less carbon-intensive industry.

Cash Payments

A third approach to assisting greenhouse gas-intensive, trade-exposed industry is a straightforward transfer of funds from the government to the companies. Such an approach offers the most flexibility in terms of how much to offer and guidance on its use. The payments could be restrictive and focused on specific issues, such as research, development, and demonstration of technologies, to more expansive concerns such as keeping the company "whole" in terms of shareholder value or other metrics. Funds for the transfer could come from either general revenues, carbon taxes, allowance auctions, or a combination of sources.

Border Adjustments: Penalize Foreign Competitors

The most direct, although potentially complex, means of addressing the trade issue would be to penalize foreign competitors who produce and export carbon intensive goods without having to meet comparable carbon policies affecting producers in the importing country. Generally with respect to climate change trade issues, the relief being sought is in the form of a border adjustment that raises the cost basis for the competing goods, potentially to a level that reflects the carbon control costs borne by the importing nation's goods.[25] Because this is in effect expanding the reach of regulation to foreign countries, implementation issues are far more complicated than the domestic-based options discussed above. The overall objective of a border adjustment would be to encourage negotiation by the United States of binding multilateral and bilateral agreements and to level the playing field with countries that have not taken action "comparable" to proposed U.S. action to reduce greenhouse gas emissions.

The two forms of border adjustments being discussed most are (1) countervailing duties, and (2) International Reserve Allowances. The primary difference between them is that the first levies conventional tariffs on imports to level the playing field, while the second imposes a shadow allowance requirement on imports to create a de facto tariff. Either could be difficult to implement. As stated by the Australian Government in its Green Paper:

> For imported goods, effective border adjustments would be very difficult to implement transparently. This is because adjustment would require accurate tracking of all inputs used in the production of a 'landed' good to determine both the amount of embedded emissions in that good and the effective carbon price that has been applied to the inputs. For example, it would be highly complex to determine the emissions and carbon cost embedded in an imported finished aluminum product. Access to reliable and robust data from other jurisdictions is not straightforward, and the complexity of the task is significantly increased when multiple jurisdictions contribute to the production of the good.[26]

Countervailing Duties

Countervailing duties are a means of providing relief to domestic industries who are subject to competition from subsidized imported products. Imposed as an additional import duty on the subsidized imported good, a countervailing duty can prevent imported goods from being sold in the domestic market at prices less than "similar" products produced domestically. Under this approach, the lack of "comparable" carbon policies by foreign countries would be considered a subsidy by the United States and a countervailing duty based on the embedded carbon in their imported good would be levied.[27] Under a carbon tax scheme, the tax would be based on the carbon tax and embedded carbon. Under a cap-and-trade program, the tax would be based on an average allowance price and embedded carbon. However, imposition of countervailing duties based on the embedded carbon in imports would raise complex issues of law under the WTO.[28]

International Reserve Allowances

In place of a countervailing duty, this second approach to providing relief is an international reserve allowance (IRA) requirement—essentially a cap-and-trade scheme focused on imports of greenhouse gas-intensive products with each IRA equal to one metric ton of carbon dioxide equivalent. Basically the IRA requirement would require that, in order for greenhouse gas- intensive products from countries with insufficient carbon policies to enter the United States, they must be accompanied by a prescribed amount of "international reserve allowances" based on greenhouse gas emissions generated in the production of the products. The import requirement would enter into effect after a reasonable time had passed for negotiations on an acceptable greenhouse gas reduction program.

Generally, proposed legislation would require the Environmental Protection Agency (EPA) to calculate an annual IRA requirement for each category of covered goods from a covered country. It would make this determination based on best available information and publish the requirements before the beginning of each compliance year. Also, EPA would be required to establish a method for calculating the required number of IRAs for each category of covered goods from a covered foreign country; the method is to apply to covered goods manufactured and processed in a single country and to require submission of IRAs on a per-unit basis for each category of covered goods from a covered country. In addition, EPA would have to establish separate procedures for determining applicable IRA requirements for goods that are "primary products" and are manufactured or processed in more than one covered country. EPA would have to revise these various IRA requirements annually.

As generally outlined in proposed legislation, a covered good under the program: (1) is a "primary product" or "manufactured item for consumption," (2) generates a "substantial quantity" of direct or indirect greenhouse gas emissions in its manufacture, and (3) is "closely related" to a good whose cost of production in the United States is affected by a requirement in the proposed legislation. A "primary product" would be iron, steel, steel mill products, aluminum, cement, glass, pulp, paper, chemicals or industrial ceramics, and any other manufactured product that is sold in bulk and generates in its manufacture direct and indirect greenhouse gas emissions comparable to emissions generated in the manufacture of products by U.S. industrial entities that would be subject to emissions caps in the proposed legislation. Indirect greenhouse gas emissions are greenhouse gas emissions resulting from the generation of electricity consumed in manufacturing of a covered good.

Comparable action is generally defined as any greenhouse gas regulatory programs, requirements and other measures that, in combination are comparable in effect to actions carried out by the United States though federal, state, and local measures to limit greenhouse gas emissions, based on best available information.

Alternative Sectoral Approaches

Sectoral approaches have become a blanket term to cover a multitude of different options to address emissions from individual industrial sectors.[29] They can be confined to domestic industries, or be international in scope. They can be voluntary or mandatory. They can be incorporated into cap-and-trade schemes, or function outside of such schemes as either an alternative reduction program (mandatory), or an exemption (voluntary). For purposes of this paper, the discussion of sectoral approaches will be limited to mandatory and voluntary schemes that address the trade impacts of carbon policies.

For example, at the Conference of Parties (COP) held at Bali in December 2007, the International Iron and Steel Institute (IISI) issued a paper calling for a carbon intensity cap on steel; the carbon intensity of steel would be determined on a per-ton basis as the embedded carbon in steel divided by its weight.[30] The foundation of this sectoral approach would be the collection of carbon dioxide data by steel plants in major steel producing countries. The data collected would be used to develop intensity-based benchmarks for the industry. IISI argues that using this comprehensive intensity-based approach to emissions from steel production "will allow production normalised CO2 emission comparisons between regions that are not possible today." As stated by IISI: "By including all the major steel producing countries, world wide competition will no longer be harmed in an industry where over 40% of products are already traded internationally."[31]

This approach has been endorsed by the American Iron and Steel Institute, whose press release states the approach is also supported by IISI members in both the developed and developing countries, including China.[32] It has also been reported by the *Financial Times* that the approach is supported by the China Iron and Steel Association.[33] Such a carbon intensity mechanism could be used for several industries for which foreign competition and emissions leakage are serious concerns. Obviously, many parameters and specifics would have to be negotiated to determine whether such an approach would be effective in addressing the concerns identified here.

The basic structure of a sectoral approach depends on the overall purpose it is designed to achieve. For example, a voluntary scheme could have the characteristics of an exemption with no penalty for failing to make progress toward achieving the overall purpose of the program. However, the program would probably only affect direct emissions; industry would still have to respond to any cost impacts from indirect emissions. Other voluntary schemes could make participation voluntary, but once a company agrees, the scheme would be mandatory for that company. This is how many approaches to encouraging participation by developing countries are designed. The current Clean Development Mechanism (CDM) reflects this line of thinking.

For a mandatory scheme, the linkages between it and the broader program would be key to accomplishing the overall purposes of the program. Because this paper limits its discussion to international approaches, the strength of a sectoral approach is that it would at least start the process of incorporating developing countries into a global approach to climate change.

Despite the wide variety of sectoral approaches proposed, there are some commonalities among them.

The first is transparency of important parameters (also called "benchmarks"), including definition and boundary of affected sectors, agreed upon performance metrics and indicators, and identification of best practices. As stated by the Centre for European Policy Studies (CEPS) Task Force:

> Without such data, collected bottom-up by industry and verified by an independent third party, there is no justification for sectoral approaches. Only verified data can ensure that industry commitments, whether voluntary, unilateral or negotiated with government, lead to 'real' and 'measurable' reductions beyond a business-as-usual scenario.[34]

The second is the sharing and dissemination of best practices within a sector to increase efficiency and transfer of technology.

The third is mechanisms to encourage incorporation of installations in developing countries into the overall scheme. Mechanisms include technical assistance, technology transfer, greenhouse gas credits for reductions, and threats of regulation.

Analysis

This section is divided into two components: (1) general issues and questions raised by the various approaches (design, effects, etc.), and (2) discussions of specific options. It is not comprehensive, but illustrative of the range of questions and consequences these initiatives present.

General Design Issues Surrounding Assistance

The design of an assistance program—the goals, eligible participants, implementation, and enforcement—would be difficult to define in a manner that satisfies all parties. There is every incentive for any industry facing a cost increase from carbon policies to claim that its competitive position could be diminished, thereby justifying special consideration by the government. The government would be in the difficult position of picking winners and losers, sometimes without access to important, but proprietary, data. The following discussion outlines some of the challenges entailed in crafting an acceptable program.

Defining Overall Goals for Assistance

At first glance, this would seem a simple question with a simple answer. However, there have been a variety of purposes and objectives suggested for proposals to assist trade-exposed, greenhouse gas-intensive industries. They include (1) promote negotiation of an international agreement; (2) prevent the leakage of carbon emissions from countries with carbon policies to those without them; (3) remove a barrier to enacting domestic legislation; (4) assist industry in making a smooth transition to a less-carbon-intensive future; (5) level the competitive playing field that carbon policies may upset; and (6) prevent or mitigate potential job losses from carbon policies. These are discussed below.

Promoting an International Agreement

As suggested earlier, the problems arising from differentiation would not exist if there were an international agreement on reducing global greenhouse gas emissions that placed all significant greenhouse gas emitting countries under a coherent regulatory regime. This criterion raises numerous questions about what would constitute a fair agreement, or comparable obligations by developing countries; however, it represents the long-term solution to the trade (and climate change) issue. In the case of sectoral approaches, this is their primary purpose, less so for domestic assistance options.

On a more practical level, the need to promote an international agreement also reflects the strictures of the WTO, if a border adjustment is being considered. Since a border adjustment may well violate the General Agreement on Tariffs and Trade (GATT), a measure that is successfully challenged on this ground would need to be justified under a GATT exception. While GATT Article XX contains an exception for "measures relating to the conservation of exhaustible resources," provided that domestic production or consumption restrictions are also imposed, such a measure may not be "applied in a manner which would constitute arbitrary or unjustifiable discrimination between countries where the same conditions prevail or a disguised restriction on international trade." In determining whether "unjustifiable" discrimination exists, the WTO Appellate Body would probably examine whether the United States had made "serious efforts" to negotiate agreements before imposing an import barrier.[35] Attempts to impose a trade barrier without such efforts would make the barrier more difficult to justify and thus more likely to be considered a WTO violation.

Preventing Carbon Leakage

The environmental rationale for seeking a sectoral agreement, or imposing trade restrictions or assisting domestic production, is to prevent carbon leakage. Such a goal is environmental, not economic. Indeed addressing carbon leakage would likely drive up the cost of compliance with any carbon reduction program. As stated by the Australian Government's *Green Paper*:

> If Australia was solely concerned about minimizing the domestic cost of meeting a reduction in emissions, it would be unconcerned about carbon leakage. However, given the global nature of the climate change problem, the potential for carbon leakage provides a rationale to use policy to influence the locational decisions of emissions-intensive industries on environmental grounds.[36]

For border adjustments, avoiding carbon leakage would also be the primary rationale for qualifying for an exception under GATT. As suggested above, the ability to separate and quantify the effects of differentiated carbon policies and the mitigating effects of any policy response would be difficult. There are no guarantees that any proposed solution would prevent carbon leakage, or that any assistance would prevent the migration of production and jobs abroad. Trade is a multi-faceted and complex series of interactions.

Preventing Job Loss

Much of the political basis for supporting sectoral approaches, domestic assistance, or trade restrictions for greenhouse gas-intensive industries is to protect domestic jobs in those industries. On a nationwide basis, greenhouse gas-intensive industries are not a substantial source of employment on a percentage basis. Data compiled by the Peterson Institute for International Economics and the World Resources Institute indicate that five such industries (ferrous and nonferrous metals, nonmetal mineral products, basic chemicals and pulp and paper) account for 1.7% of U.S. employment (2.25 million jobs) and 3.0% of Gross Domestic Product (GDP).[37] Overall, manufacturing is responsible for about 10.6% of U.S. employment and 12.4% of GDP. In addition, a carbon policy is likely to create jobs in other parts of the economy, such as renewable and energy conservation technologies, reducing or potentially eliminating job loss on a nationwide basis.

Nevertheless, this would be cold comfort to communities directly affected by potential job loss from the potential trade imbalance created by differentiated carbon policy. Although a small percentage of total employment and GDP, factories and companies can be a significant employer and generator of wealth in some states and local communities. For example, the manufacturing sector in Indiana produces 18.9% of the state's payroll jobs and 30.2% of its GDP.[38]

Although job loss is a major concern, it is not a concern recognized by the WTO as a rationale to justify a GATT-inconsistent measure. In addition, any assistance provided to industry does not guarantee that jobs will not be lost or moved. As suggested previously, locational decisions by companies are multi-faceted: Assistance to mitigate the effects of carbon policies will not necessarily affect competitiveness issues with respect to labor rates, exchange rates, or other relevant factors.

Leveling the Competitive Playing Field

This is the most publicized economic argument in favor of sector approaches, domestic assistance, or tariffs for greenhouse-gas intensive, trade-exposed industries. However, this is a somewhat vague notion as it is not clear what, or how much, assistance would level the playing field, and at what costs to other parts of the economy. As stated by the Australian Government:

> It is difficult to determine how much EITE [emissions-intensive trade-exposed] assistance would be needed to prevent carbon leakage. Some have argued that there is a direct relationship between a loss in profitability and carbon leakage, and that Government intervention could be warranted to restore the profitability of EITE entities to levels that would have occurred without a carbon constraint. In the extreme case, and all other things constant, this would imply assistance at a direct dollar-for-dollar rate for the impact of the carbon price. Under such an approach, the Government would continue to provide assistance even if other factors substantially increased the profitability of EITE entities. ...
>
> The level of assistance to EITE industries over time must also be balanced against the impact on non-assisted sectors. In particular, the design of the EITE-assistance policy needs to take into account the fact that a declining national emission cap combined with a growing national economy implies that the burden (or cost) of achieving a given national reduction in emissions would increase over time. This suggests that the degree of EITE assistance may need to be adjusted over time to ensure the sustainability of the EITE policy, otherwise EITE assistance would constitute a growing share of a shrinking quantity of national emissions, leading to higher costs for the rest of the economy.[39]

This need for balance in any assistance is echoed by other studies. The Peterson Institute for International Economics and the World Resources Institute suggest that focusing on the competitive concern of carbon intensive industries is a "fairly narrow interpretation of U.S. competitiveness."[40] Following the rationale of the Australian Government's *Green Paper*, the two institutions make three arguments for caution in designing assistance for greenhouse gas-intensive industries: (1) a move to a more carbon-constrained economy requires a "fundamental" shift that requires a strong regulatory environment to promote; (2) assistance to greenhouse gas intensive, trade-exposed industries comes at a cost to the economy as a whole; (3) to the extent assistance to greenhouse gas intensive, trade-exposed industries delays reductions by those industries, that delay has costs in terms of increased emissions.[41]

Encouraging a Smooth Transition to a Less-Carbon-Intensive Future

If one accepts the need for a transition to a less-carbon-intensive future, assistance to greenhouse gas-intensive, trade-exposed industries could be justified to the extent it promotes such a transition with less economic pain. As stated by the Australian Government:

> The second reason [after avoiding carbon leakage] for assisting trade-exposed industries is that it may smooth the transition of the economy towards one that embodies a price on

> carbon. Given the significant differences between the emissions profiles of industries, a carbon price could have a markedly greater impact on some industries than on others. Government could place a priority on providing transitional assistance to those entities and industries that would be most severely affected by the introduction of the scheme. This would involve giving priority towards assisting existing industries, particularly those with significant "sunk" capital investments, few opportunities to reduce their emissions profiles and a limited capacity to pass through the carbon cost.[42]

This purpose reinforces the need for a balance between the desire to "level the playing field" as suggested above, with the need to achieve the overall environmental goal that the carbon policy is designed to achieve.

Remove a Barrier to Domestic Legislation

This purpose reflects the historic difficulties in the United States of committing to national, mandatory emission targets.[43] The reluctance of the United States to adopt mandatory actions reflects concerns about costs, as witnessed by the U.S. negotiation and ratification of the 1992 United Nations Framework Convention on Climate Change (UNFCCC).[44] The UNFCCC reflects this negotiating position of the United States and some other countries in that it calls for voluntary control measures. Senate floor debate on ratification of the treaty brought out concerns by some Senators about the *cost* of compliance, its impact on the country's *competitiveness*, and the *comprehensiveness* with respect to the developing countries—concerns that were overcome because of the non-binding nature of the reduction goals.[45]

Assistance to greenhouse gas intensive, trade-exposed industries is a direct attempt to respond to the competitiveness and comprehensiveness concerns that have been expressed in Congress and other venues for almost two decades.

Defining Eligible Industries

Among the fundamental questions any assistance program must answer is "Who is eligible for assistance?" The previous discussion suggests that greenhouse gas-intensive, trade-exposed industries could be the most competitively disadvantaged by carbon policies that increase costs. Those increased costs could be imposed directly if the reduction program included industrial emissions under its regime, and/or indirectly through increases in immediate products those industries consume in the making of their products (such as energy). Three criteria stand out for determining the potential eligibility of an industry, sub-industry, or company for assistance.

- Is the sector or product greenhouse gas-intensive? Under some metric (profit, value added, employment), is the sector's viability strongly tied to a greenhouse gas-intensive process?
- Would the sector's competitive situation be substantially upset by carbon policies through its inability to pass on costs related to them? Under a fragmented international regime, would the sector be exposed to competition from companies in countries not anticipated to respond for some time with significant carbon policies of their own?

- Does the sector have only a limited ability to cost-effectively reduce its emissions or obtain compliance through another means, at least in the short-term? Does this situation present a significant downside risk economically in terms of lost production and jobs and environmentally in terms of carbon leakage?

What the steel example discussed earlier suggests is that products that are greenhouse gas- intensive with relatively low value added (such as crude steel) are potentially most at risk of significant cost increases. If, in addition, high greenhouse gas-intensive, low added-value products are fairly homogeneous and can be readily bought in international markets, domestic manufacturers of them (such as raw steel producers) may be price-takers on world markets with limited ability to pass through carbon-related costs. Parts of several industry sectors may fall into this category, including cement, lime, some basic chemicals, and primary metals (such as primary aluminum), along with some glass and paper products. However, the steel example illustrates that defining "greenhouse gas-intensive, trade-exposed" industries will not be a straightforward process, as trade exposure, carbon costs, and pricing dynamics may differ within a sector. Under some assistance options, developing eligibility criteria could put the government in the position of picking winners and losers, and creating the potential for a drawn-out and litigious process.

Overall, the discussion suggests that determining industry eligibility would not be straightforward and would require drawing lines and making fine distinctions. Issues requiring resolution would include (1) the level of disaggregation to use in determining eligibility; (2) the metric that would be used to determine eligibility; and (3) the entity and data that would do the determining. All of these determinations would be controversial.

Implementation Issues

It is the details of the proposed assistance that would determine its effectiveness in achieving any of the purposes discussed above. Some of the more critical implementation questions are identified below.

Allocation —How Much?

The above discussion suggests an important tradeoff to the economy between providing assistance to greenhouse gas-intensive, trade-exposed industries and increasing the burden of carbon policy compliance on other parts of the economy. Assistance would increase the overall cost of compliance in hope of achieving a smoother economic transition to a low-carbon future. However, determining the appropriate amount would be controversial and contentious. Issues include:

- How much should be allocated and in what form (tariffs, domestic assistance, sectoral approach, etc.)?
- How much should the assistance be tailored to individual sectors, subsectors, or facilities?
- What metrics and baselines should be used to make these decisions?

Duration of Assistance — How Long?

If a long-term, substantial greenhouse gas reduction is desired, all sectors would have to participate in the reductions at some point. The longer participation by one sector is delayed, the higher the costs on the participating sectors. However, the ability of greenhouse gas-intensive, trade-exposed industries to join a reduction program may vary substantially, depending on research and development results, compliance strategies by industries providing important feedstocks to their processes, and general economic conditions and demand for their products. Issues include:

- How differentiated should the timing of the assistance be by sector, subsector, or facility?
- How should any adjustments to assistance over time be determined?
- Under what conditions should the assistance be terminated?

Implementation and Enforcement — How Will it Work?

Each option faces significant implementation and enforcement issues. Fragmentation is a key characteristic of greenhouse gas intensive, trade-exposed industries, in terms of their greenhouse gas intensiveness, their trade exposure, and their sector economics. This situation would put substantial demands on the implementing body with respect to data needs and methodologies. Data and methodologies would need to be robust enough to justify determinations (including any WTO challenges in the case of border adjustments), and to enforce any requirements on domestic or foreign producers. Finally, enforcement would require some definition of success or failure. Issues raised include:

- How will the necessary data be collected and quality assured?
- How will international cooperation be encouraged, both to negotiate an agreement and to implement any tariff?
- What metrics and methodologies will be used and how will they be tested for rigor?
- What criteria will be used to determine success or failure?

Data Needs

The data needs for all these options are substantial—particularly for trade and sectoral approaches. At the current time, there would be a clear tradeoff between the precision of a trade or sectoral approach and the ability of the government to implement it. The international scope of these two approaches multiplies the data challenges presented by at least an order of magnitude over a domestic-only program. The challenge for data collection in developing countries may be such that the government is forced to employ methodology, rather than empiricism, to construct "data" sets—a process that would make a WTO challenge (in the case of trade approaches) almost a certainty. A sectoral approach may be able to solicit assistance from those countries, if they feel the approach is fair to them and that improved efficiency and technology will improve their economic situation. In contrast, the sanctions approach of the trade schemes may not encourage such countries to cooperate in the scheme.

Potential for Unintended Consequences

Attempting to resolve an international problem – the lack of a comprehensive international climate change treaty – unilaterally can be an uncertain enterprise. The approaches outlined here face daunting needs in terms of crafting a coherent program to achieve multiple goals. There is a high probability of unintended consequences from any of these approaches. Trade and economics involve dynamic processes that can respond to public policy in unanticipated ways. For example, trade sanctions based on primary goods, such as steel and aluminum, could have undesirable impacts on domestic downstream industries. An increase in the cost of raw steel or aluminum could drive up the costs of domestically manufactured finished products, such as automobiles, and encourage foreign countries to export more finished products to the United States. Indeed, a country could redirect its exports from primary goods to finished goods to avoid the trade sanctions. For example, South Korea, which exports both raw steel and automobiles, could focus its industrial policy toward automobile exports and away from raw steel exports. Thus, downstream companies that use greenhouse gas-intensive goods could have their competitiveness undermined by attempts to protect greenhouse gas-intensive, trade-exposed industries. This consequence is less likely with domestic assistance or an international sectoral approach.

Another potential unintended consequence of a trade approach is that foreign countries with more stringent carbon polices than those proposed in the United States could turn the tables. The European Union (EU) has already agreed to a more stringent reduction program to 2020 than the United States seems likely to adopt. Even if a U.S. trade program did not target the EU (because of the "comparable" provisions), it is conceivable that the EU might target the United States.

There is also a risk that domestic subsidies could lead to unintended outcomes. For example, a company receiving assistance might choose to use that money for something other than modernizing or operating targeted carbon-intensive facilities. Instead, it might decide that the overall competitiveness of a plant does not merit any modernization, and choose to close the facility or reduce its production regardless of any assistance.

Issues for Specific Approaches

Free Allowance Allocation

Free allocations of allowances to greenhouse gas-intensive, trade-exposed industry is the most popular means of assistance for countries under the Kyoto Protocol. For phases 1 and 2 of the European Union's (EU) Emissions Trading Scheme (ETS), member countries have almost exclusively allocated allowances at no cost, and over-allocated in favor of industries in competitive markets, compared with the electric power sector. Likewise, the Australian Government's *Green Paper* recommends free allocation of allowances under its proposed capand-trade program to assist greenhouse gas-intensive, trade-exposed industries. Finally, New Zealand has announced that it intends to use free allocation as its means of assisting its industries.

A primary advantage of a free allocation system is that it doesn't necessarily exempt greenhouse gas-intensive, trade-exposed industries from the cap-and trade program. Thus, cost-effective reductions may still be made, lowering overall cost of the program compared with an approach that exempts them completely. Also, incorporating their emissions in the cap from the beginning helps industries become familiar with the workings of the carbon market and how to develop least-cost strategies to comply with increasingly stringent reductions and likely reductions in free allocations. This could assist in determining how long any assistance should be in effect and with the smooth transition objective identified above. Also, as the EU-ETS experience suggests, attempting to add exempted industries in a piecemeal process can be a difficult task. Putting them under the cap from the beginning makes the direction of greenhouse gas emission policy for industry clear.

This is not to say that designing a free allocation system would be simple. As suggested previously, there are at least two major points of contention in the design of such an approach: (1) What percentage of the total available allowances should be allocated free to greenhouse gas- intensive, trade-exposed industries? and (2) What methodology and metrics should be used to apportion the free allowances among the various industries and sub-industries? The first point of contention highlights the zero-sum game that is allowance allocations under a cap-and-trade program: allowances given free to greenhouse gas-intensive, trade-exposed industries cannot be given to other heavily impacted industries (such as electric utilities) or sold by the government at auction to fund other government objectives or tax reform. Resolution of this tradeoff would determine how much relief greenhouse gas-intensive, trade-exposed industries would receive. For example, the Australian's Government's green paper recommends up to 30% of available allowances be allocated free to greenhouse gas-intensive, traded-exposed industries. This would be allocated under a two-tier system where heavily greenhouse gas-intensive industries (on a revenue basis) would receive free allowances to cover 90% of their emissions, while somewhat lesser greenhouse gas-intensive industries would receive 60%.[46]

A variety of metrics and options are available for resolving the second point of contention. Free allowance allocations could be weighted in a manner to encourage increased domestic production of greenhouse gas-intensive, trade-exposed goods and to discourage "off-shoring" of that production (e.g., an output based allocation). Such a methodology would help meet objectives such as reduced carbon leakage and reduced job losses. Another example would be to protect shareholder value: keeping the companies "whole." In its green paper, the Australian Government recommends a metric based on greenhouse gas emissions per unit of revenue, stating: "A measure of emissions per unit of revenue would be the most transparent and comparable indicator of the materiality of the carbon cost impact across different traded industries."[47]

From an economic standpoint, an important disadvantage of free allocation is that allowances allocated free are allowances the government cannot auction and from which there are no proceeds to address other concerns. Economic studies have found that, if revenues received from an auction-based allocation system are used in the most economically efficient manner, the overall costs imposed on the economy by a cap-and-trade program could be reduced substantially.[48] Economists maintain that the most economically efficient application of auction revenues would be as an offset for reductions made in taxes on desirable activities, such as employment or personal income.[49] Likewise, the auction revenues could be used to support other public polices, such as research and development or relief to low-income families.[50] To the extent allowances are allocated free to greenhouse gas-intensive, trade-exposed industries, these other options are excluded.

A second disadvantage is that the difficulty in determining an appropriate apportionment of allowances opens the possibility of "windfall profits" by some industries or sub-industries. The amount necessary to compensate industries varies by industry.[51] In addition, the allowance price will also vary over time, and may not strictly track costs. Thus, the chances that some industries could be over-compensated is significant.[52] This issue has been raised with the EU-ETS and is of continuing concern there.[53]

A final disadvantage of free allocation is that it might not achieve some of the purposes outlined above. Particularly if allowances are apportioned according to historic production, companies may simply pocket the allowances and still lower production or move off-shore. This can be avoided if the apportionment is based on output. Likewise, the option may have no effect on the crafting of an acceptable international agreement.

Carbon Tax Credits

Under a carbon tax scheme, carbon tax credits to greenhouse gas-intensive, trade-exposed industries would be similar, in effect, to a free allocation under a cap-and-trade system. The primary advantage of a carbon tax credit option is that most carbon tax proposals assume real-time emissions monitoring (or derivative calculations based on real-time fuel consumption). Real- time allocation (as opposed to allotments based on historical emissions) would naturally respond to changes in production (and derivative job gains/losses). There would be little chance for "windfall" profits, assuming accurate emissions monitoring.

The primary disadvantage of a carbon tax credit is that, depending on how it is designed, greenhouse gas-intensive, trade-exposed industries receiving the credits could have little incentive to make greenhouse gas reductions on their own. The precision available in allocating tax credits removes carbon price as a factor in production and planning. This problem could be addressed by a phase-out schedule tailored to encourage commercialization of more carbon- efficient technology and processes, although trying to tailor such a phase-out to each industry or sub-industry's specific situation could be complicated and contentious.

Cash Payments

The primary advantage of cash payments is transparency. Particularly with free allowance allocations under a cap-and-trade program, there is some veil over exactly how much is being given to individual sectors, subsectors, or companies. With cash payments, it can be made clear who is getting what and how much. The allocations could be a matter of public record, making public and congressional oversight more straightforward. Likewise, any phasing out of the assistance over time would be clear.

The disadvantage of cash payments is possible imprecision in allocations: Under a cap-and trade program, cash payments would be based on estimates of allowance prices and would have to be reconciled with the actual price at some point (such as the end of the year). That estimates would be off is likely, providing affected facilities with either a short term, no-cost loan (if too high), or a short-term added expense (if too low). More precision is likely under a carbon tax regime. However, arguably, a cash payment program would be redundant under a carbon tax scheme as a tax credit could also be transparent and little would be gained by having the government collect the money only to return it (particularly in the case of direct emissions).

Border Adjustments[54]

The two versions of border adjustments identified in this chapter have been discussed extensively. Countervailing duties have been widely discussed in Europe, with an eye on imposing such a requirement on the United States,[55] and an international reserve allowance scheme has been embodied in several legislative proposals in the United States. Border adjustments are seen as being relatively economically efficient (compared with domestic assistance). The duty on imports would allow the domestic carbon program to be implemented in the most cost-effective manner without the distorting effects of targeted, domestic assistance, while protecting greenhouse gas- intensive, trade-exposed industries from being unfairly targeted by foreign competitors not undergoing a transition to a less-carbon intensive economy.

Whether it would level the playing field with respect to international trade, or encourage foreign countries to pass carbon policies of their own is more debatable. For example, avoiding carbon leakage with a trade approach may neither encourage major developing countries like China and India to commit to carbon targets, nor greatly influence their overall exports. As suggested by American Enterprise Institute (AEI) Center for Regulatory and Market Studies in a recent report:

> As a means of coercing China, this strategy would face long odds. First, why would China and India, by adopting domestic GHG controls, handicap all of their global trade merely to avoid sanctions on a quite small part of their economies? Less than 1 percent of Chinese steel production is sold to America in a form that would make it liable to sanctions. For aluminum, the number is only 3 percent. It is 2 percent for paper and less than 1 percent for both basic chemicals and cement. Second, one country adopting trade sanctions, or a few countries doing so, will merely change the geographic pattern of trade flows without having much impact on the total demand for Chinese energy-intensive goods. U.S. sanctions on China would cause countries with low-carbon steel, aluminum, or other industries to increase their exports to the U.S. and increase their own imports from China. It is implausible to suggest that this threat would compel China to adopt GHG controls that would remotely resemble the severity of those being proposed in America.[footnote omitted][56]

The international scope of the border adjustment approach and the complex nature of trade makes design of a program difficult. Terms including "comparable action," "similar products," and "embedded carbon" would have to be defined in a manner that avoids arbitrary and unjustifiable discrimination between exporting countries in order to comply with WTO requirements, and methodologies developed to give meaning to them. Then a price per ton of embedded carbon would have to be determined. Assuming a carbon tax scheme domestically, this price would be obvious; in an International Reserve Allowance system under a cap-and-trade scheme, the price would have to be linked in some manner to prevailing allowance prices.

The definitions and methodologies needed to implement a border adjustment have different webs of complexity. For example:

- How close should "comparable" action be to "identical"? If a country achieves the same percentage reduction without any comprehensive program, is that "comparable"? What if that program achieves the same results as the United States, but exempts greenhouse gas-intensive, trade-exposed industries? If a country achieves more stringent reduction levels than the United States, does the United States concede the right for them to impose a border adjustment against it? When does a comparable action have to occur? How should country of origin be determined?

- How is "similar product" determined? To what category, sub-category, or product-specific level will determinations be made? How broad should the scope of the program be? How will necessary international trade and carbon data be collected to determine appropriate baskets of covered products? How do WTO requirements affect the basket of covered products?
- How is "embedded carbon" determined? At what level of aggregation will the determination be made? Can a country have excessive embedded carbon for one product, but not another? Can different companies within a country have different embedded carbon estimates? What about products whose manufacture involves several countries, some with comparable policies and some without? Will products produced by different processes be considered together or separately? What about the same products made with different energy sources? What baselines should be used? How is potential gaming of the system prevented?

In addition, an International Reserve Allowance (IRA) requirement raises various implementation issues surrounding the need to administer a separate cap-and-trade program for IRAs. Beyond the operating mechanics of a cap-and-trade program, the government would have to develop a pricing mechanism for IRAs, and a compensating mechanism to account for any allowances allocated free to domestic producers. Depending on the scope of the IRA scheme, its cap-and-trade system could be substantial.

Sectoral Approaches

Sectoral approaches have been suggested by various parties, such as the steel industry's proposal noted earlier, and were added to the negotiating agenda at the Bali conference of parties as "cooperative sectoral approaches and sector-specific actions." Because they cover a wide range of options, the following discusses them in terms of basic issues, such as measuring success, financing mechanisms, and crediting mechanisms.

Measuring Success

Several metrics could be used to determine an appropriate scheme for industry. These metrics are not mutually exclusive; different industries could employ different metrics depending on the specifics of that sector's processes.

The most common metric being discussed for a sectoral-based approach is an output-based performance standard. An output-based performance standard measures success by the amount of greenhouse gases emitted per unit of output. Also called a carbon intensity target, this approach does not limit total emissions (like an emissions cap), and, therefore, is seen by its proponents as being more acceptable to developing countries who may see an emissions cap as restricting their right to development.

Obviously, a performance standard requires an agreed-upon standard, or benchmark, for participating companies to achieve. A benchmark allows participants to compare their performance against an industry standard, optimal technology, or best practice. Benchmarks can be developed at different levels of aggregation and by different methods (e.g., technical assessment, historical averages, negotiation). The most precise and effective benchmarks are based on technical assessment of best available technologies or practices, designed at a micro level, and take into account the specific products and input mixes at a plant level. Thus plants with similar processes, products, and inputs can compare their performance with each other and identify needed improvements. To the extent benchmarking[57] is used on a more aggregated level (including different processes, all similar products, etc.) the ability to improve individual plants or processes may be lost.[58]

In a strictly voluntary scheme, benchmarking could be a means to determine best practice and target technical assistance. However, it would not necessarily result in reductions if other factors, such as low energy prices, make achieving best practice not cost effective, and the company refused to join the effort. In addition, while technical-based benchmarks are a useful tool to determine the current status of best practices within an industry, it provides little guidance on the speed and magnitude of future technological advancements. Under a strictly voluntary system, achieving the benchmark could become the end of the effort—an end short of the ultimate goals of the program.

In a mandatory system (or a voluntary system that becomes mandatory upon acceptance), carrots and sticks could be used to encourage industry to move toward the benchmark. Work by Vanderborght, Baron, et al., provides one illustration of how this might work.[59] As indicated by **Figure 5**, companies operating at carbon intensity levels above the present industry average (pink line that could be determined globally or differentiated by country) would be quickly and increasingly penalized (red area), while companies operating below the average would receive a modest and declining reward (blue area). Based on assessments of future advancements, the baseline could be extended in the long-term. Additional carrots could be made available to companies that adopt innovative processes that achieve this long-term goal (green area). The appropriate sticks and carrots would depend on how the program is integrated into the overall program and are discussed under "crediting mechanisms."

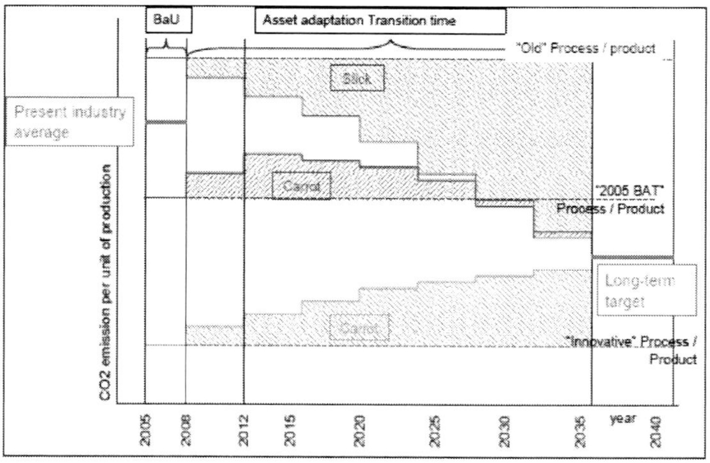

Source: Figure was modified by Baron, op. cit., from original presentation by B. Vanderborght, *The Cement-EU ETS Kaleidoscope* (2006), Presentation at the WBSCD-IEA Cement Workshop, Paris, September 5.

Figure 5. Differentiated Benchmarking: Illustration of Incentives and Evolution

However, technical benchmarks, whether for a voluntary or a mandatory program, have drawbacks, most notably around their need for data. Five identified by Baron et al. are as follows:

- Benchmarking is a time-consuming, data-intensive activity, all the more so as various conditions may need to be accounted for in an international approach. There is a risk of inflation in the number of benchmarks, as operators will argue special circumstances that all require special treatment.
- In some cases, benchmarking may require disclosing data that companies judge proprietary or of strategic importance. This may be handled through a careful choice of performance indicators used in the benchmark.
- It is a useful tool to describe an industry status "here and now" but as it is based on today's technologies and practice, it provides little guidance on what level mitigation can be achieved in the future – as in some cases, technology is yet to be invented. Can a benchmark then be used as a forward looking method?
- The use of an average industry benchmark as a reference to allocate effort will immediately define "winners" and losers" – i.e., installations that perform better or worse than the chosen benchmark target. While the effect on their cost would be a fair reflection of the cost associated with CO2 emissions, it may be difficult to agree to, unless the benchmark is set as a future target, as illustrated in Figure 5.
- There is an asymmetry of information between any industry and a government when it comes to assessing the ability to adjust processes and to invest in new technologies to reduce greenhouse gas emissions. It is not, *a priori*, in an industrial actor's interest to reveal the full extent of its mitigation potential and its real cost.

Because of these data needs, other possible performance standards have been proposed, such as industrial average carbon intensity. Over time, the standard would be strengthened at an agreed- upon rate toward the most efficient company or companies. A trading program could be created between companies, with companies with intensities greater than the industry average buying necessary allowances from companies with intensities less than the industry average. Essentially, the system would constitute a separate cap-and-trade program based on carbon intensity rates rather than a cap based on annual emissions. While such an approach would mitigate some of the data needs of a technically based benchmark, issues such as level of aggregation for averages would remain.

Financing Mechanism

Because a major focus of a sectoral approach is to encourage carbon policies in developing countries, a financing mechanism to assist the transfer of technology and expertise is usually included in a proposal, particularly the more voluntary the approach is.[60] These mechanisms may also include assistance to domestic companies that are inefficient producers to bring them up to the agreed-upon performance standard. For example, the Dutch domestic sectoral approach to improve industrial energy efficiencies includes financial and regulatory incentives to encourage industry to sign voluntary agreements to reduce energy intensity.[61] The funding source for these incentives could come from allowance auctions (under a cap-and-trade program), carbon tax revenues, or other mechanisms.

Crediting Mechanism

Most proposals to credit reductions under a sectoral approach are tied to the allowances used in a cap-and-trade program. As noted earlier, allowances are essentially a form of currency that can be converted to a monetary value via a market. In some ways, a voluntary sectoral scheme with crediting mechanisms already exists under the Kyoto Protocol: The Clean Development Mechanism (CDM) provides a means for financing and receiving credit for installing technology that reduces emissions in countries without mandatory carbon policies.[62]

However, CDM projects are on an ad hoc basis, and a sectoral scheme designed to address the purposes identified here would require more structure and direction. As illustrated by Baron and Ellis in Figure 6, the project-by-project baseline and reduction target of CDM would need to be replaced by a broader country-specific, or policy specific, baseline and a calculated emission reduction target. The approach would cover the entire sector, not just the most cost-effective opportunities as with CDM. Facilities that perform poorly would diminish the total quantity of credits available to the sector as a whole, unless the program was voluntary with a "no-lose" provision.[63] The difficulties in setting these baselines and reduction targets were noted previously.

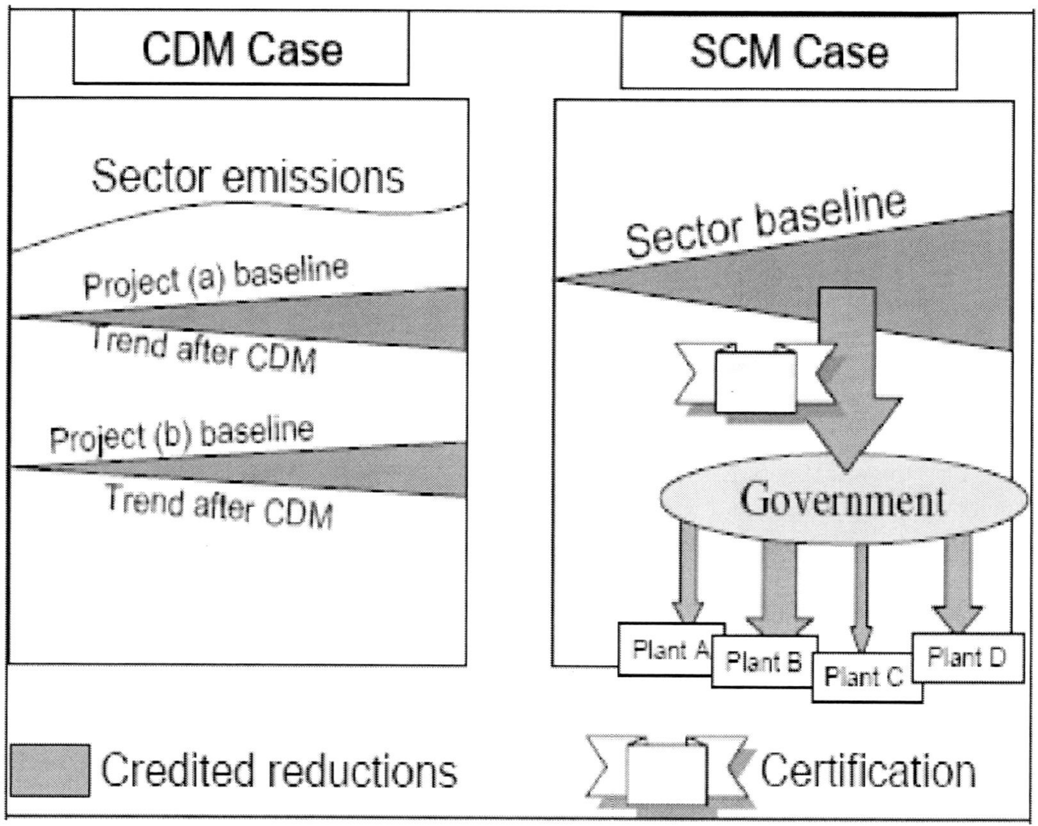

Source: Richard Baron et al., *Sectoral Approaches to Greenhouse Gas Mitigation* (November 2007), p. 32.

Figure 6. Clean Development Mechanism (CDM) vs. Sectoral Crediting Mechanism (SCM)

Table 5. Summary of Major Approaches to Carbon Leakage and Competitiveness Concerns

Approaches	Promoting International Agreement	Prevent Carbon Leakage	Prevent Job Loss	Level the Playing field	Encourage Smooth Transition	Remove Barrier to Legislation
Assist Domestic Industries through free allowance allocations, tax credits, or cash payments	Not a primary focus of the approach	Depends on what industry does with the allocations, tax credit, or cash payment	Depends on what industry does with the allocations, tax credit, or cash payment	Amount of leveling depends on design details: who is included, what emissions are included (direct, indirect), and amount of assistance provided	Effectiveness depends on design details, including phase-out chedule and support for new technology	Free allocation of allowances has been the dominant option for addressing competitive issues in Europe, proposed for Australia and New Zealand
Penalize foreign competitors through countervailing duties or an international reserve allowance program	Depends on how trading partners respond to the coercive nature of trade barrier	Depends on how trading partners respond to the trade barrier. Trading partners may shift from covered primary goods to downstream finished goods with no effect on carbon leakage	Depends on the reaction of the entire economy to the tariffs—jobs saved in covered industries could result in jobs lost in downstream industries	Depending on the metric chosen, playing field may be returned to what it was before the imposition of carbon policies for chosen industries; may disturb the playing field for downstream industries	Effectiveness depending on design details, including phase-out schedule and support for new technology	International reserve allowance schemes have been incorporated in leading cap-and-trade legislation in the United States
Incorporate industries through a sectoral scheme	Primary focus of the approach	Depends on substitution effects and how voluntary it is	Depends on substitution effects and how voluntary it is	Assuming global coverage and participation, the playing field could become significantly more even across countries depending on the performance metric chosen	Depends on metric used for performance benchmark and schedule for efficiency improvement	Sectoral schemes have received trade associations' endorsements and are recognized in the Bali Action Plan

IMPLICATIONS

Table 5 summarizes the three general approaches to address trade-related issues with respect to the various objectives of these approaches. As indicated, each focuses on different objectives. With respect to achieving the more comprehensive solution to trade issues—promoting an international agreement—the range presented by the three approaches is clear and distinct. Support for domestic industries, the approach most commonly included in legislative proposals, is not focused on this objective; it is focused on preserving the industry's current competitive position and jobs and may, depending on the details, help transition that industry to the future. Trade measures for foreign competitors, another approach commonly included in legislative proposals, may provide a stick for international negotiation, but the primary focus is on protecting greenhouse gas-intensive, trade-exposed industries from "unfair" competition while the country awaits an international agreement. Finally, the sectoral approach represents a range of options focused on integrating developing countries' industrial bases into a mutually acceptable international framework that provides a level playing field for all participants. Whether any of these approaches would have appreciable effects on carbon leakage is unclear.

As the U.S. debate on climate change proceeds, various proposals for reducing greenhouse gas emissions contain provisions to address the trade-related issues presented here. Two of the most common options are (1) subsidies for affected industries through allocation of free allowances within cap-and-trade policies; and (2) border adjustments through an international reserve allowance program. In addition to these domestic options, the Bali Action Plan includes sectoral approaches as options for the next phase of the Kyoto Protocol. Other alternatives are also likely to be debated.

Free allocation of allowances to greenhouse-gas intensive, trade-exposed industries is more narrowly focused on assisting domestic industries maintain their current competitiveness in the face of a domestic greenhouse gas reduction program. It has the virtue of relative simplicity compared with the other approaches and options, but no greater guarantee of success. Companies may choose to accept the assistance and not make the necessary improvements to existing facilities to remain competitive in the increasingly carbon-constrained future. There are metrics and benchmarks that could be used to allocate free allowances (carbon intensity, output-based metrics) that can reduce these problems, but introduce complexity in terms of increased data needs and methodological considerations. The allocation also comes at the cost of making the program more expensive for the other participants.

The International Reserve Allowance scheme is a complex system focused on leveling the playing field for domestic producers against competitors whose countries are not implementing comparable greenhouse gas reductions. The international scope of the option and the complex nature of trade makes design of a program difficult. Terms including "comparable action," "similar products, and "embedded carbon" would have to be defined in a manner that avoids arbitrary and unjustifiable discrimination between exporting countries in order to comply with WTO requirements, and methodologies developed to give meaning to them. Annual assessments of countries' actions would have to be made and new baselines set. Then a price per ton of embedded carbon would have to be determined. The use of allowances, instead of money, removes the transparency of a countervailing duty and makes tracking the impact of the scheme on other parts of the economy difficult to determine. Finally, it is unclear how the affected trading partners would respond, economically, environmentally, or politically.

This might encourage one to move to the sectoral approach as potentially more effective. It is focused on achieving an international agreement that would make the playing field at least acceptable to all parties. However, there is no blueprint currently that parties agree is the basis for developing such an approach. It is possible that the 2009 Copenhagen conference will resolve the fundamentals for such an approach, but what sort of contingencies one should consider in the meanwhile, or in the face of failure, is unclear.

Finally, it is the details of any of these options that would ultimately determine their effectiveness in achieving the various objectives. The potential options are almost endless. For example, if a domestic assistance approach is chosen, allocation options include production output, historic emissions, company profits or revenues, and technology or best practices benchmarks. Duration options for such an approach would include anticipated technology or best practices advancements (or best currently available), consummation of an international agreement, or some criterion related to the economic health of the sector or industry. Options for data collection include publicly available data from the Departments of Energy and/or Commerce, legislatively- mandated requirement for the collection of data from companies wishing to receive assistance, or government estimates based on best available data or modeling. In any case, the task would be daunting for any of the approaches.

The design of an assistance program—the goals, eligible participants, implementation, and enforcement—would be difficult to define in a manner that satisfies all parties. There is every incentive for any industry facing a cost increase from carbon policies to claim that its competitive position could be diminished, thereby justifying special consideration by the government. The government would be in the difficult position of picking winners and losers, sometimes without access to important but proprietary data.

End Notes

[1] As an expression of these concerns, during deliberations on the Fiscal 2009 Budget Resolution (S.Con.Res. 70), the Senate agreed to a motion to instruct conferees "that no legislation providing for new mandates on greenhouse gas emissions should be enacted until it effectively addresses imports from China, India, and other nations that have no similar emissions programs." Motion agreed to by a vote of 55-40 (Senate Roll Call Vote 132, May 15, 2008).

[2] For a further discussion of the evolution of U.S. climate change policy, see CRS Report RL30024, *U.S. Global Climate Change Policy: Evolving Views on Cost, Competitiveness, and Comprehensiveness*, by Larry Parker and John Blodgett.

[3] The green paper defines Emissions-intensive, trade exposed industries as: "Industries that either are exporters or compete against imports (traded exposed) and produce significant emissions in their production of good (emissions intensive)." Department of Climate Change, Commonwealth of Australia, *Carbon Pollution Reduction Scheme: Green Paper* (July 2008), p. 60.

[4] Department of Climate Change, Commonwealth of Australia, *Carbon Pollution Reduction Scheme: Green Paper* (July 2008), p. 292.

[5] For a broader definition of carbon leakage, see Glen P. Peters and Edgar G. Hertwich, "Trading Kyoto," *Nature Reports: Climate Change* (April 2008), pp. 40-41.

[6] U.S. Environmental Protection Agency, Office of Atmospheric Program, *EPA Analysis of Lieberman-Warner Climate Security Act of 2008: S. 2191 in the 110th Congress* (May 14, 2008), pp. 81-85, 165-167.

[7] For a further discussion of the uncertainties involve in estimating the effects of climate change legislation over 40 years, see CRS Report RL34489, *Climate Change: Costs and Benefits of S. 2191/S. 3036*, by Larry Parker and Brent D. Yacobucci.

[8] Non-Annex I countries are ones not having set carbon reduction goals. They are countries not listed in Annex I to the U.N. Framework Convention on Climate Change (1992); the Annex I nations are designated, developed nations that agreed to reduce emissions.

[9] U.S. Environmental Protection Agency, Office of Atmospheric Program, *EPA Analysis of Lieberman-Warner Climate Security Act of 2008: S. 2191 in the 110th Congress* (May 14, 2008), p. 84.

[10] For a review of these studies, see Julia Renaud, *Issues Behind Competitiveness and Carbon Leakage: Focus on Heavy Industry*, (OCED, October 2008).

[11] Terry Barker and Igor Bashmakov, *IPCC WG3 2007, Chapter 11: Mitigation From a Cross-Sectoral Perspective* (March 6, 2008), presentation by Michael Grubb, p. 26.

[12] Department of Climate Change, Commonwealth of Australia, *Carbon Pollution Reduction Scheme: Green Paper* (July 2008), p. 292.

[13] Carbon Trust, *The European Emissions Trading Scheme: Implications for Industrial Competitiveness* (June 2004), pp. 6-7.

[14] Department of Climate Change, Commonwealth of Australia, *Carbon Pollution Reduction Scheme: Green Paper* (July 2008), pp. 208-309.

[15] Kasten Neuhoff and Susanne Droege, *International Strategies to Address Competitiveness Concerns* (July 2007), p. 4.

[16] Kasten Neuhoff and Susanne Droege, *International Strategies to Address Competitiveness Concerns* (July 2007), p. 4.

[17] Kasten Neuhoff and Susanne Droege, *International Strategies to Address Competitiveness Concerns* (July 2007), p. 4.

[18] Carbon Trust, *The European Emissions Trading Scheme: Implications for Industrial Competitiveness* (June 2004), p. 6.

[19] Misato Sato and Karsten Neuhoff, *Testing the Demand Substitution Effect* (June 22, 2007).

[20] The HHI is calculated by squaring the market share of each firm competing in the market and then summing the resulting numbers.

[21] Peterson Institute for International Economics and World Resources Institute, *Leveling the Carbon Playing Field: International Competition and US Climate Policy Design* (2008), p. 8.

[22] McKinsey & Company, Reducing U.S. Greenhouse Gas Emissions: How Much at What Cost? (December 2007).

[23] McKinsey & Company, Reducing U.S. Greenhouse Gas Emissions: How Much at What Cost? (December 2007), p. 51.

[24] Carbon Playing Field: International Competition and US Climate Policy Design (2008), p. 8.

[25] For background on trade remedies, see CRS Report RL32371, *Trade Remedies: A Primer*, by Vivian C. Jones.

[26] Department of Climate Change, Commonwealth of Australia, *Carbon Pollution Reduction Scheme: Green Paper* (July 2008), p. 300.

[27] For more information on this approach, see Joseph E. Stiglitz, *Making Globalization Work*, (New York, 2006).

[28] Robert Howse, "Book Review: The End of the Globalization Debate: A Review Essay," *Harvard Law Review*, (2007-2008) p. 1542.

[29] For a review of options, see Richard Baron et al., *Sectoral Approaches to Greenhouse Gas Mitigation* (November 2007).

[30] International Iron and Steel Institute, A Global Section Approach to CO2 Emissions Reduction for the Steel Industry (December 2007).

[31] International Iron and Steel Institute, A Global Section Approach to CO2 Emissions Reduction for the Steel Industry (December 2007), p. 4.

[32] American Iron and Steel Institute, North American Steelmakers Endorse Global Sectoral Approach to Addressing Climate Change (December 13, 2007).

[33] Peter March, "China Trade Body Backs Check on Steel Emissions," *Financial Times* (October 10, 2007).

[34] Bjorn Stigson, Christian Egenhofer and Noriko Fujiwara, *Global Sectoral Industry Approaches to Climate Change: The Way Forward* (Brussels, 2008) p. 24.

[35] For example, see Appellate Body Report, *United States Import Prohibition of Certain Shrimp and Shrimp Products* (October 10, 1998), at paras. 166-172.

[36] Department of Climate Change, Commonwealth of Australia, *Carbon Pollution Reduction Scheme: Green Paper* (July 2008), p. 293.

[37] Peterson Institute for International Economics and World Resources Institute, *Leveling the Carbon Playing Field: International Competition and US Climate Policy Design* (2008), p. 11.

[38] Jerry N. Conover, "Indiana," *Indiana Business Review*, Outlook 2008 (2008), pp. 11-12.

[39] Department of Climate Change, Commonwealth of Australia, *Carbon Pollution Reduction Scheme: Green Paper* (July 2008), pp. 294, 296.

[40] Peterson Institute for International Economics and World Resources Institute, *Leveling the Carbon Playing Field: International Competition and US Climate Policy Design* (2008), p. 10.

[41] Peterson Institute for International Economics and World Resources Institute, *Leveling the Carbon Playing Field: International Competition and US Climate Policy Design* (2008), pp. 11-12.

[42] Department of Climate Change, Commonwealth of Australia, *Carbon Pollution Reduction Scheme: Green Paper* (July 2008), p. 294.

[43] For further information, see CRS Report RL30024, *U.S. Global Climate Change Policy: Evolving Views on Cost, Competitiveness, and Comprehensiveness*, by Larry Parker and John Blodgett.

[44] The United States signed the UNFCCC on June 12, 1992, and ratified it on October 15, 1992. The UNFCCC entered into force on March 21, 1994. For a review of the negotiations, see CRS Report 92-374, *Earth Summit Summary: United Nations Conference on Environment and Development (UNCED), Brazil, 1992*, by Susan R. Fletcher.

[45] *Congressional Record*, Vol. 138 (October 7, 1992), 33520-33527.

[46] Department of Climate Change, Commonwealth of Australia, *Carbon Pollution Reduction Scheme: Green Paper* (July 2008), pp. 319-321.

[47] Department of Climate Change, Commonwealth of Australia, *Carbon Pollution Reduction Scheme: Green Paper* (July 2008), p. 311.

[48] For example, see Lawrence H. Goulder, *Mitigating the Adverse Impacts of CO2 Abatement Policies on Energy-Intensive Industries* (2002), Resources for the Future Discussion Paper.

[49] See Lawrence H. Goulder, Mitigating the Adverse Impacts of CO2 Abatement Policies on Energy-Intensive Industries (2002), Resources for the Future Discussion Paper; Anne E. Smith and Martin T. Ross, Allowance Allocation: Who Wins and Loses Under a Carbon Dioxide Control Program? (2002), Charles River Associates; Dallas Burtraw, et. al., The Effect of Allowance Allocation on the Cost of Carbon Emission Trading (2001), Resources for the Future.

[50] For more on auctions, see CRS Report RL34502, *Emission Allowance Allocation in a Cap-and-Trade Program: Options and Considerations*, by Jonathan L. Ramseur.

[51] Richard D. Morgenstern, et. al., *Competitiveness Impacts of Carbon Dioxide Pricing Policies on Manufacturing* (November 2007), Resources for the Future, Issue Brief 7, p. 105.

[52] Allowances could also be allocated under a cap-and-trade program through technology-based benchmarks. It could also be used to allocate any allowances to new entrants.

[53] See CRS Report RL34150, *Climate Change and the EU Emissions Trading Scheme (ETS): Kyoto and Beyond*, by Larry Parker.

[54] Countervailing duties and an international reserve allowance scheme present many of the same issues. Thus, they are generally discussed together here, with differences noted as appropriate.

[55] The debate on this has been heated at times. For example, see AFP, *Climate Change: Sarkozy backs carbon tax, EU levy on non-Kyoto imports* (October 25, 2007), available at [http://afp.google ALeqM5gx9Wyuo7XJiydxsqseJmVdX3-MoQ]. For a review of proposals, see Julia Renaud, *Issues Behind Competitiveness and Carbon Leakage: Focus on Heavy Industry*, (OCED, October 2008) pp. 77-79.

[56] Lee Lane and David Montgomery, *Political Institutions and Greenhouse Gas Controls*, (November 2008) p. 11.

[57] Benchmarking can be defined as "the process of identifying the best practice in relation to products and processes, both within an industry and outside it, with the object of using it as a guide and reference point for improving the practice of one's own organization." "Benchmarking," *Dictionary of Business*, (Oxford University Press, 2002).

[58] Richard Baron, et al., *Sectoral Approaches to Greenhouse Gas Mitigation* (November 2007), pp. 28-29.

[59] Figure was modified by Baron, op. cit., from original presentation by B. Vanderborght, *The Cement-EU ETS Kaleidoscope* (2006), Presentation at the WBSCD-IEA Cement Workshop, Paris, September 5.

[60] See Jake Schmidt, et al., Sector-based Approach to the Post-2012 Climate Change Policy Architecture, (August 2006), pp 13-20; and, Julia Renaud, Issues Behind Competitiveness and Carbon Leakage: Focus on Heavy Industry, (OCED, October 2008) p. 100.

[61] See Valentina Dinica, Hans Th. A. Bressers, and Theo de Bruijn, "The Implementation of a multi-annual agreement for energy efficiency in The Netherlands," *Energy Policy* (2007) pp. 1196-1212; and Nicole van Deeck, Transferring the Dutch Success of Voluntary Agreements with Industry to other Countries," The European Council for an Energy- Efficient Economy, Summer 2007 Study (June 2007), panel 7.

[62] For a discussion of CDM as a crediting mechanism for sectoral approaches, see Julia Renaud, *Issues Behind Competitiveness and Carbon Leakage: Focus on Heavy Industry*, (OCED, October 2008) pp. 97-100.

[63] Richard Baron et al., *Sectoral Approaches to Greenhouse Gas Mitigation* (November 2007), pp. 31-32. For information on "no lose" sectoral approaches, see Jake Schmidt et al., *Sector-based Approach to the Post-2012 Climate Change Policy Architecture* (August 2006).

In: Carbon Leakage: Trade Issues in Emission Reduction ISBN: 978-1-61728-919-4
Editor: Victoria T. Greenwald

Chapter 2

CLIMATE CHANGE: EUROPEAN UNION AND PROPOSED UNITED STATES APPROACHES TO CARBON LEAKAGE AND WTO IMPLICATIONS

Larry Parker and Jeanne J. Grimmet

SUMMARY

The United States has proposed, and the European Union (EU) developed, policies to mitigate the potential economic and environmental (i.e., "carbon leakage") impacts of carbon policies on energy- or greenhouse gas-intensive, trade-exposed industries. While studies have found little effect of carbon policies on EU competitiveness in the present, the EU decision to move toward auctioning of allowances in the future has spurred development of criteria to extend potential availability of free allowances to exposed industries to 2020. In a December 2009 decision, the European Commission (EC) listed 164 industrial sectors and subsectors deemed exposed sectors under appropriate European Parliament and Council directives.

H.R. 2454, which passed the House on June 26, 2009, includes two strategies to address these concerns: (1) free allocation of allowances (similar to that of the EU), and (2) an international reserve allowance (IRA) scheme. Studies have suggested that a free allowance scheme appears effective in mitigating the trade-related impact of the carbon program on energy-intensive, trade- exposed industries. However, production cost for those industries (along with other industries) could increase because of the potential pass-through of compliance-related costs by upstream producers of various inputs into their manufacturing processes. Whether these costs would become significant would depend on the ability of upstream suppliers to pass on the costs, and the ability of the downstream industries to respond by increasing the efficiency of their operations or by substituting other, less-costly inputs into their processes. There are questions about whether the allowances provided by H.R. 2454's allocation scheme are sufficient. If the Environmental Protection Agency's estimates are correct, the allocation would appear sufficient. If industry estimates are correct, or if individual showings of eligibility prove significant, the pool of allowances provided by the bill would appear inadequate under the assumptions used here. Also, the data and administrative resources necessary to implement the program would be substantial.

Although H.R. 2454 as passed would require EPA to establish an IRA program consistent with U.S. international agreements, questions may be raised as to whether proposed Part IV and its application would fully comply with U.S. international trade obligations. The distribution of free allowances may constitute actionable subsidies for purposes of the World Trade Organization (WTO) Agreement on Subsidies and Countervailing Measures by possibly qualifying as "foregone revenue" when auctioning of allowances would also be permitted. In addition, the requirement that importers purchase IRAs to accompany particular imports might be found to constitute a prohibited import surcharge or, if the product may not otherwise enter the United States, a prohibited quantitative restriction under the General Agreement on Tariffs and Trade (GATT) 1994. While the IRA program might be provisionally justified under GATT general exceptions for health protection or resource conservation, the GATT also requires that it not be applied "in a manner that would constitute a means of arbitrary or unjustifiable discrimination between countries where the same conditions prevail, or a disguised restriction on international trade." Whether an IRA program can be applied consistently with these requirements may depend on the type of program that may be crafted by EPA under the proposed legislation—that is, on the elements that would be required under the bill and the administrative possibilities inherent in its discretionary authorities. Absent an international consensus on the types of trade-related measures that may be applied as part of a domestic climate change regime, adversely affected countries may seek to challenge these measures under WTO dispute settlement provisions. Since neither the distribution of emission allowances nor border restrictions imposed as part of a domestic greenhouse gas-reduction program have yet come before WTO dispute settlement panels, WTO obligations and exceptions remain untested in this complex regulatory environment.

INTRODUCTION

Congress is considering legislation to reduce emissions of greenhouse gases that may, depending on the specifics of the final legislation, affect the competitiveness of energy- or greenhouse gas- intensive industries. Competitiveness can be a rather abstract term for which any precise meaning can be elusive.[1] Competitiveness is a continuing phenomenon, with companies becoming more or less competitive according to a host of factors, including productivity, market demand, resource costs, labor costs, exchange rates, and the like. As stated by the Australian Government in its Green Paper on carbon reduction schemes:

> Changes in the cost structures of entities and industries are not unusual and occur continuously in a market-based economy; nor is it unusual for Government policy to change cost structures. For example, the adoption of high quality occupational health and safety standards have affected the profitability of Australia's labour-intensive traded industries, making it more difficult for them to compete with foreign producers that are subject to lower standards. Assistance is not usually provided to offset the impact of domestic policies on traded industries, as those policies reflect the priorities and values of the Government and community more generally.[2]

Most industries face a competitive market (sometimes international in scope) both in terms of producers of the same products and producers of substitute products. Also, in some cases, an industry may face a fairly elastic demand for its product. Thus, most industries are price sensitive, and therefore any increase in manufacturing costs—as by a carbon emission reduction requirement—hurts the competitiveness of a firm. This complex situation is further complicated for energy-intensive industries as competitors within the same industry may experience different energy price increases (particularly for electric power), depending on their individual energy needs and power arrangements. For example, an aluminum plant receiving power from a hydroelectric facility may not be affected the same way as a similar plant whose power contract is with a coal-fired power supplier.

The addition of a carbon control regime to this competitive dynamic has raised concerns that, in the absence of similar policies among competing nations, if the United States adopts a carbon control policy, energy- or greenhouse gas-intensive, trade-exposed industries that must control their emissions or that find their feedstock or energy bills rising because of costs passed-through by suppliers may be less competitive and may lose global market share (and jobs) to competitors in countries lacking comparable carbon policies. In addition, this potential shift in production could result in some of the U.S. carbon reductions being undercut by increased production in less regulated countries; this is commonly known as "carbon leakage."

Greenhouse gas reduction legislation introduced over the last two Congresses has included provisions to address carbon leakage and to mitigate the effect of carbon policies on U.S. competitiveness. In general, two strategies have been proposed: (1) providing assistance to greenhouse gas-intensive, trade-exposed industries; and (2) imposing tariffs on certain greenhouse gas-intensive goods imported into the country from countries not implementing comparable carbon policies. Such tariffs are frequently referred to as border measures. H.R. 2454, as passed by the House, contains both of these strategies.[3]

This chapter examines the dynamics of this issue in three parts. First, the European Union (EU) has been implementing a cap-and-trade program for four years, and has finalized a third reduction phase that will run from 2013 through 2020.[4] This chapter reviews and analyzes the experience of the EU in addressing its concerns about energy-intensive, trade-exposed industries, and the lessons those efforts may have for the United States. Second, the House-passed American Clean Energy and Security Act of 2009 (H.R. 2454) contains both a free allocation scheme and a border measure among its provisions to address the concerns of energy-intensive, trade-exposed industries. This chapter reviews and analyzes these provisions. Third, these same provisions could come under scrutiny under various U.S. trade agreements, particularly within the World Trade Organization (WTO). Concerns have been expressed that the border measure contained in H.R. 2454 would be suspect under various provisos of the WTO. This chapter analyzes the potential WTO implications of any attempt to implement a subsidy or a border measure under H.R. 2454.

USING FREE ALLOCATIONS UNDER THE EU-ETS: RESULTS AND LESSONS LEARNED

Background on the European Union's Emissions Trading Scheme (ETS)

The EU's Emissions Trading System (ETS) covers more than 10,000 energy-intensive facilities across the 27 EU Member countries, including oil refineries, powerplants over 20 megawatts (MW) in capacity, coke ovens, and iron and steel plants, along with cement, glass, lime, brick, ceramics, and pulp and paper installations. In addition, aviation is currently being phased into the ETS. These covered entities emit about 40%-45% of the EU's total greenhouse gas emissions, and almost two-thirds of them are combustion installations. The trading program does not cover either carbon dioxide (CO_2) emissions from the transportation sector (except aviation), which account for about 25% of the EU's total greenhouse gas emissions, or emissions of non-CO_2 greenhouse gases, which account for about 20% of the EU's total greenhouse gas emissions. A Phase 1 trading period ran between January 1, 2005, and December 31, 2007.[5] A Phase 2 trading period began January 1, 2008, covering the period of the Kyoto Protocol, and a Phase 3 has been finalized to begin in 2013.[6]

Under the Kyoto Protocol, the then-existing 15 nations of the EU agreed to reduce their aggregate annual average emissions for 2008-2012 by 8% from the Protocol's baseline level (mostly 1990 levels) under a collective arrangement called a "bubble." In light of the Kyoto Protocol targets, the EU adopted a directive establishing the EU-ETS that entered into force October 13, 2003.[7] One objective of the second phase of the ETS is to achieve 3.3 percentage points of the 8.0% reduction required by the EU- 15 under the Protocol.[8]

The importance of emissions trading was elevated by the accession of 12 additional central and eastern European countries to EU membership from May 2004 through January 2007. For the new EU-27, the overall ETS emissions cap is set at 2.08 billion metric tons of carbon dioxide (CO_2) annually for the Kyoto compliance period (2008-2012).

The second phase Kyoto compliance stage of the ETS is built on the experience the EU gained from its preliminary Phase 1. The European Commission (EC) believes that the Phase 1 "learning by doing" exercise prepared the community for the difficult task of achieving the reduction requirements of the Kyoto Protocol. Several positive results from the Phase 1 experience assisted the ETS in making the Phase 2 process run smoothly, at least so far. First, Phase 1 established much of the critical infrastructure necessary for a functional emission market, including emissions monitoring, registries, and inventories. Much of the publicized difficulties the ETS experienced early in the first phase can be traced to inadequate emissions data infrastructure.[9] Phase 1 significantly improved those critical elements in preparation for Phase 2 implementation.

Second, the ETS helped jump-start the project-based mechanisms—Clean Development Mechanism (CDM) and Joint Implementation (JI)—created under the Kyoto Protocol.[10] As stated by Ellerman and Buchner:

> The access to external credits provided by the Linking Directive has had an invigorating effect on the CDM and more generally on CO_2 reduction projects in developing countries, especially in China and India, the two major countries that will eventually have to become part of a global climate regime if there is to be one.[11]

Third, according to the EC, a key result of Phase 1 was its effect on corporate behavior. An EC survey of stakeholders indicated that many participants are incorporating the value of allowances in making decisions, particularly in the electric utility sector, where 70% of firms stated they were pricing the value of allowances into their daily operations, and 87% into future marginal pricing decisions. All industries stated that it was a factor in long-term decision-making. [12]

European Energy-Intensive, Trade-Exposed Industries

Background

Figure 1 below indicates the cost sensitivity of various manufacturing activities in the United Kingdom as determined by Climate Strategies.[13] Cost sensitivity is measured as the percentage of the activity's current gross value added at stake from a 20 euro per metric ton carbon price. [14] As indicated by the **bold**, several of these industries are covered by the ETS, including lime, cement, basic iron and steel, refined petroleum products, pulp, paper and paperboard, hollow glass, and flat glass. The figure also indicates the direct and indirect cost components of implementing a carbon pricing policy. The cost impact of a 20 euro carbon price from a manufacturing process from direct emissions is indicated by the light blue versus the cost impact of indirect emissions resulting from higher electricity prices, which is indicated by the dark blue. As shown, the balance of direct and indirect costs differs substantially among the various sectors.

EC Phase 3 Decision on Eligible Industries

After nine eastern European Member States threatened to veto an initial proposal to auction 100% of all allowances, the leaders of the European Union (EU) agreed to provide for some free allocation of allowances during Phase 3 that will begin in 2013.[15] In making changes for Phase 3, the European Commission has identified three CO_2 emitting sectors for inclusion under the ETS: petrochemicals, ammonia, and aluminum.[16] The ETS would also expand beyond CO_2 to include nitrous oxide (N_2O) emissions from nitric, adipic, and glyoxalic acid production, and perfluorocarbon (PFC) emissions from the aluminum sector. These industries would be added to those currently covered: oil refineries, powerplants over 20 MW in capacity, coke ovens, and iron and steel plants, along with cement, glass, lime, brick, ceramics, and pulp and paper installations (aviation is currently being incorporated into the system).

Most covered industries, except for electric powerplants, will be eligible for some free allocation of allowances to cover direct emissions under the Phase 3 agreement. For electric powerplants, most will receive no free allocation of allowances during Phase 3. However, in a concession to certain eastern European Member States, an optional and temporary derogation from the no-freeallocation requirement for powerplants is provided to countries that meet specific energy and economic criteria. Under the optional allocation scheme, the Member State can allocate allowances equal to 70% of the powerplant's Phase 1 emissions free; this allocation will decline in the out-years.

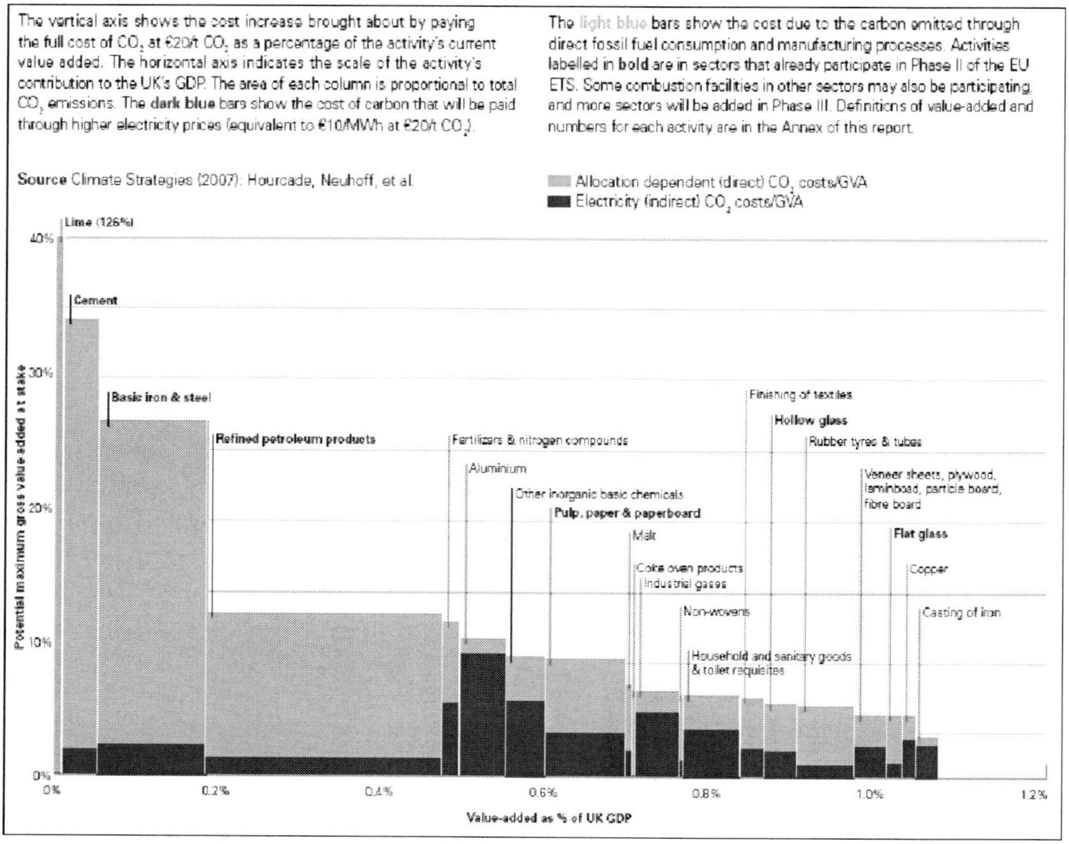

Source: Carbon Trust, *EU ETS Impacts on Profitability and Trade* (January 2008), Based on data in Climate Strategies (2007).

Notes: Annex referenced in Figure 1 can be found on page 32 of source report.

Figure 1. UK Manufacturing Activities Most Cost-sensitive to CO_2 Pricing

The auction schedule for most other covered entities is more gradual with 80% of a sector's allocation provided free in 2013, declining linearly to 30% by 2020, and zero by 2027. As stated in the final EC directive:

> For other sectors covered by the Community scheme, a transitional system should be foreseen for which free allocation in 2013 would be 80% of the amount that corresponded to the percentage of the overall Community-wide emissions throughout the period 2005 to 2007 that those installations emitted as a proportion of the annual Community-wide total quantity of allowances. Thereafter, the free allocation should decrease each year by equal amounts resulting in 30% free allocation in 2020, with a view to reaching no free allocation in 2027.[17]

For energy-intensive, trade-exposed industries, Phase 3 has provisions to provide assistance to eligible installations to address the direct and indirect impact of emissions control costs. With respect to direct emissions costs, the EC published a list of installations exposed to a significant risk of carbon leakage on December 24, 2009.[18] The list is identical to the draft list released in September 2009.[19] The decision lists 164 industrial sectors and subsectors deemed exposed sectors under the appropriate European Parliament (EP) and Council directives. That list is provided in the Appendix. Eligible installations will receive allowances sufficient to cover 100% of their direct emissions, provided they are using the most efficient technology available. Reflecting the fluid nature of the competitive situation and international negotiations, the EC is to review its decision by June 30, 2010, and provide the EP and Council with any appropriate proposals to respond to the situation.

Assistance for the impact of indirect emissions control costs on exposed industries would be determined by Member States. As stated in the final directive:

> Member States may deem it necessary to compensate temporarily certain installations which have been determined to be exposed to a significant risk of carbon leakage related to greenhouse gas emissions passed on in electricity prices for these costs. Such support should only be granted where it is necessary and proportionate and should ensure that the Community scheme incentives to save energy and to stimulate a shift in demand from grey to green electricity are maintained.[20]

ANALYSIS OF EU APPROACH

Effectiveness of Phase 1 Free Allowance Allocations

In general, allowances have been allocated free to participating entities under the ETS. During Phase 1, the EU-ETS Directive allowed countries to auction up to 5% of allowance allocations, rising to 10% under Phase 2.[21] Under Phase 1, only four of 25 countries used auctions at all, and only Denmark auctioned the full 5%. The political difficulty in instituting significant auctioning into ETS allowance allocations is the almost universal agreement by covered entities in favor of free allocation of allowances and opposition to auctions.[22] Free allocation of allowances represents a one-time transfer of wealth to the entities receiving them from the government issuing them.[23] The resulting transfer of wealth has been described by several analysts as "windfall profits."[24] As summarized by Ellerman and Buchner: "Allocation in the EU-ETS provides one more example that, notwithstanding the advice of economists, the free allocation of allowances is not to be easily set aside."[25]

Despite concerns about windfall profits and economic distortions resulting from the free allocation of allowances, there is little change in basic allocation philosophy for Phase 2. No country proposed auctioning the maximum percentage of allowances allowed (10%). Most do not include auctions at all.[26] The unwillingness of governments to employ auctions as an allocating mechanism revolves around equity considerations, including (1) the inability of some covered entities to pass through cost because of regulation or exposure to international competition; (2) the potential drag on a sector's economic performance from the up-front cost of auctioned allowances; and (3) the potential that government will not recycle revenues to alleviate compliance costs, international competitiveness impacts, or other equity concerns, resulting in the auction costs being the same as a tax.[27]

Most studies of the competitiveness impacts of the ETS during Phase 1 have found no impact. The International Energy Agency (IEA) cites several reasons for this situation:

> Experience to date with the EU-ETS does not reveal leakage for the sectors concerned—analysis of steel, cement, aluminum and refineries sectors reveals that no significant changes in trade flows and production patterns were evident during the first phase (2005-2007) of the EU-ETS. This is mostly due to the free allocation of allowances, sometimes in generous quantities, and to the still functioning long-term electricity contracts, which softened the blow of rising electricity prices. Further, the general boom in prices for most traded products subject to carbon costs—whether direct or indirect—has blurred any effects of the latter. Finally, the relatively short time span of these policies does not allow observation of the full potential effects on industry via changes in investment location decisions.[28]

This conclusion is echoed by Carbon Trust, which states that currently, free allocation of emissions allowances offset almost all of the additional costs of the ETS; and it is echoed by The Climate Group for The German Marshall Fund, which states that companies surveyed found it difficult to quantify effects on their bottom line in the first phase, or found no effect at all.[29]

Current Attitude of Companies under ETS

As noted earlier, the EC believes that one of the major positive outcomes of the ETS has been the incorporation of carbon prices in EU corporate decision-making. A survey of EU-ETS companies by Point Carbon suggests this assertion is true.[30] As indicated in **Figure 2**, companies are factoring the long-term price of carbon into their future investment decisions. According to Point Carbon, it is the power sector and the pulp and paper sectors that appear to consider the carbon price most decisive in their planning.

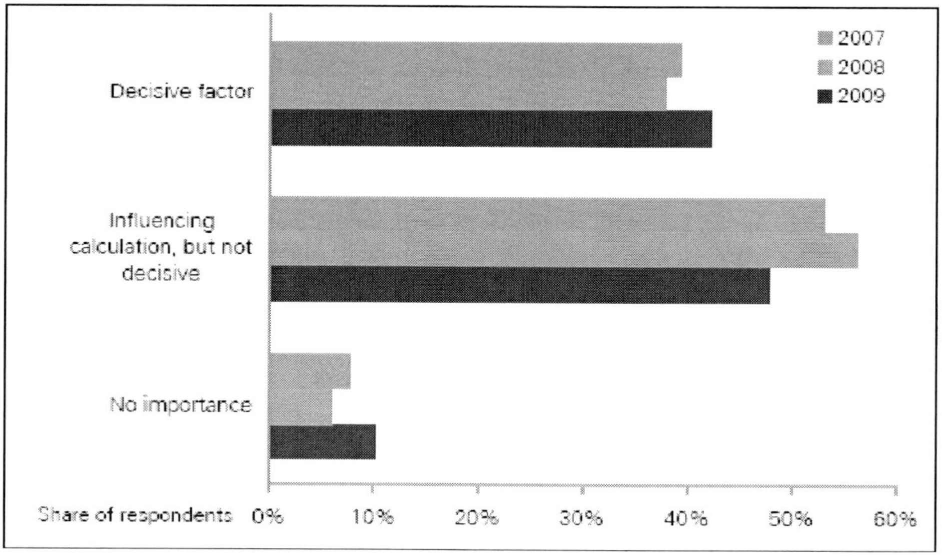

Source: Point Carbon, *Carbon 2009*, p. 10.

Notes: Long-term defined as 2020 in the questionnaire. A total of 301 companies affected by the EU-ETS were surveyed by Point Carbon.

Figure 2. How Important Is the Long-Term Carbon Price for New Investments in Your Industry?

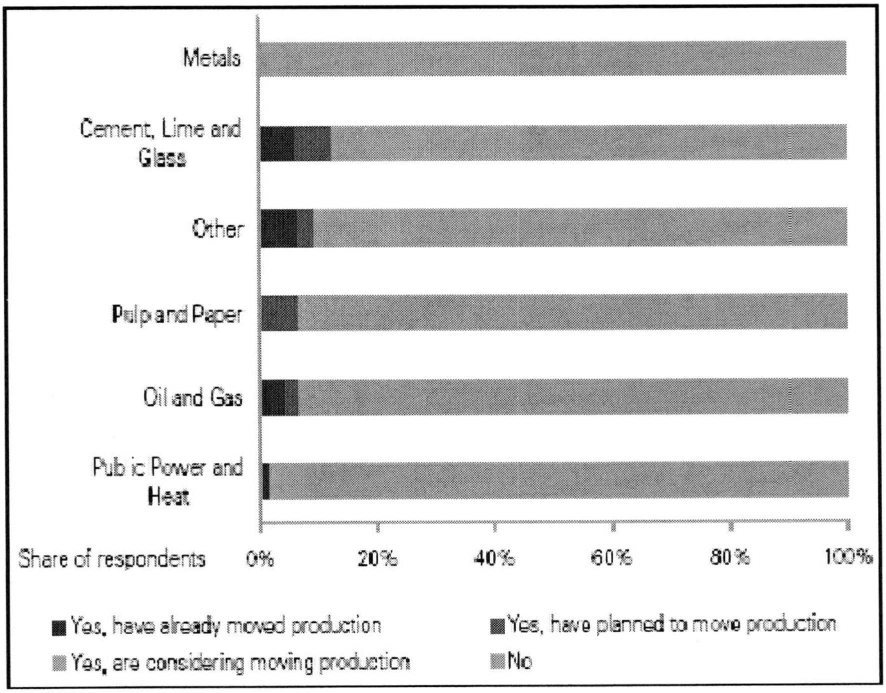

Source: Point Carbon, *Carbon 2009*, p. 12.
Notes: A total of 301 companies affected by the EU-ETS were surveyed by Point Carbon.

Figure 3. Companies Response to Carbon Price

With respect to considering moving production to other countries because of carbon prices, the Point Carbon survey of EU-ETS companies does not reveal a major trend yet. As indicated in Figure 3, over 80% of companies surveyed have not considered moving production because of carbon pricing; however, some of that includes companies, like power producers, that have limited relocation opportunities. A more detailed look at the figure indicates that 44% of the respondents in the metals, cement, lime, and glass sectors have at least thought about moving production. This may be one reason the EU included provisions extending the free allocation of allowances to such energy-intensive, trade-exposed sectors through Phase 3.

These findings by Point Carbon were generally confirmed by the survey and analysis conducted by The Climate Group for The German Marshall Fund.[31] Among that survey's conclusions were the following:

- Although costs for some firms are increasing, there is scant evidence of effects on competitiveness—but concerns about the future persist, especially as the number of free allowances decreases and CO_2 costs are reflected in electricity prices. The survey noted that aluminum smelters were particularly sensitive to electricity costs and that the pass-through of CO_2 costs may affect future production decisions.
- Companies have not relocated their operations, reduced their workforce, or lost market share as a result of carbon pricing to date.

- A market price for carbon has, to date, had a relatively low impact on how top management runs their businesses. But companies are quick at internalizing the EU-ETS into their strategic planning. Short-, medium-, and long-term effects of carbon pricing on strategic planning vary.

Determining Eligibility for Phase 3: The EC's List

Criteria Used for EC Eligibility List

Analysis indicates that there are industries that could be significantly impacted by the advent of higher carbon prices under Phase 3. As indicated by Figure 4, the cement, steel, and aluminum industries are considered by Carbon Trust to be the industries most exposed to higher carbon prices. As indicated above, the EU policies being developed are to buy time for these industries with free allowances while negotiating an international response that would level the playing field for all companies within a sector.

The list created by the EC suggests a more comprehensive view of potentially affected industries than that suggested above. The EC used five different sets of criteria in compiling its list of 164 subsectors and sectors.

Significantly: Cement/clinker; steel from blast oxygen furnaces; aluminium	EU cement and steel producers could lose up to 8% market share to overseas production in central price cases with highest trade sensitivities. Sufficient free allocation to maintain their profits can buy time to negotiate a multilateral response to trade exposure.
Plausibly: Fertilisers & nitrogen compounds; 'other' inorganic basic chemicals; pulp, paper and paperboard	Should be in the EU ETS with a compensating rate of free allocation, combined with other measures to help them tackle their exposure to carbon and electricity costs.
Possibly, at higher CO_2 prices: Some refineries; manufacture of glass; household paper; tyres; copper; potentially one or two other basic chemical processes	At higher carbon prices, some products from some refineries and a few other big activities could face trade impacts. Should be in the EU ETS; modest free allocation in Phase III, particularly for new sectors, would protect profits and give time to invest in lower carbon solutions, but should not extend beyond that.
Significantly, but very small activities: Notably lime production	Loss of market share to overseas production would involve tiny absolute carbon leakage. A political decision as to whether to ignore, offer protection, or exempt.

Source: Carbon Trust, *EU ETS Impacts on Profitability and Trade* (2008), p. 3.
Note: 159 manufacturing activities studied.

Figure 4. Carbon Trust Assessment of Exposure to Phase 3 Competitiveness Issues (for the United Kingdom manufacturing sector)

1. Paragraph 4: Significant risk of carbon leakage criteria based on a sector's or subsector 's ability to pass on the direct and indirect costs of control and allowance costs into its product's price without a significant loss of market share to less carbon efficient installations outside the Community (in accordance with Article 1 0a(14) of Directive 2003/87).
2. Paragraph 5(a): Significant risk of carbon leakage criteria based on whether a sector's or subsector 's direct and indirect costs of control and allowance costs would represent a substantial increase of production costs, calculated as a proportion of the gross value added, of at least 5% and the intensity of trade with third countries, defined as the ratio between the total value of exports to third countries plus the value of imports from third countries and the total market size for the Community (annual turnover plus total imports from third countries), is above 10% (in accordance with Article 1 0a(1 5) of Directive 2003/87BC).
3. Paragraph 5(b): Significant risk of carbon leakage criteria based on whether a sector's or subsector 's direct and indirect costs of control and allowance costs would represent a particularly high increase of production costs, calculated as a proportion of the gross value added, of at least 30% (in accordance with Article 1 0a(1 6) of Directive 2003).
4. Paragraph 5(c): Significant risk of carbon leakage criteria based on a sector's or subsector 's intensity of trade with third countries, defined as the ratio between the total value of exports to third countries plus the value of imports from third countries and total market size for the Community (annual turnover plus total imports from third countries), is above 30% (in accordance with Article 10a(16) of Directive 2003).
5. Paragraph 14: Significant risk of carbon leakage criteria based on a qualitative assessment of a sector or subsector; criteria may include increased production costs, current and projected market characteristics, and profit margins (in accordance with Article 1 0a(1 7) of Directive 2003/87/EC).

A company's ability to compete under a carbon policy depends on three primary factors: (1) the greenhouse gas intensity of a company's products, which influences the company's profitability and the products' cost; (2) the company's ability to pass on any increased costs to consumers without losing market share or profitability; and (3) the company's ability to mitigate carbon emissions, reducing the impact of the carbon policy on its operations and profitability.[32] Interestingly, only the second set of criteria used by the EC seems to incorporate all three factors in determining eligibility. Indeed, the fourth set of criteria used by the EC is based solely on trade-exposure: the impact of carbon control is not included in the criteria. The expansive eligibility requirements under the third and fourth sets of criteria results in 117 of the 164 sectors and subsectors listed by the EC, and includes everything from the manufacturing of wines to numerous textiles. While such sectors are trade-exposed, they are not generally considered to be greenhouse-gas intensive, as indicated by the Carbon Trust analysis cited above.

Data Sources Used for EC Eligibility List

The EC's inclusion of a qualitative set of eligibility criteria is suggestive of the data difficulties in setting up a comprehensive program to address carbon leakage. The EC's discussion of a qualitative analysis of the "Finishing of textiles" sector (Nomenclature des Activites Economiques or NACE code 1730) presented in paragraph 17 of the draft decision is illustrative:

> A qualitative assessment has been carried out on the sector of "Finishing of textiles" (NACE code 1730), primarily due to the fact that no official trade data at the Community level is available to assess trade intensity and that all other textile sectors are highly trade intensive. The assessment demonstrated increased international competitive pressure, significant drop in production in the Community over the last years and negative or only very modest profit margins for the years evaluated, which limit the capacity of installations to invest and reduce emissions. Based on the combined impacts of those factors, the sector should be deemed as exposed to a significant risk of carbon leakage. (paragraph 17)

The "Finishing of textiles" sector is not covered by the ETS and, therefore, the EC's concern about the sector's "very modest profit margins" that "limit the capacity of installations to invest and reduce emissions" seems somewhat irrelevant, at least at the current time. The sector may be trade-exposed; however, its primary impact from controlling greenhouse gas emissions is the indirect effects of increased electricity generating costs. As indicated by Figure 1, this sector's indirect emissions account for less than 5% of its gross added value at stake. In its analysis of affected industries, Carbon Trust notes that the UK textiles finishing sector is not trade-exposed; "By far the most economically significant activities [within the UK textile industry], textiles finishing (at 230 million pounds GVA) appears in the source data as trading only domestically. Therefore, no major activities appear subject to significant carbon price impacts."[33]

Data difficulties expand beyond determining eligibility of domestic industry sectors or subsectors. As noted in paragraph 22 of the draft decision, the list is supposed to take into account the extent to which third countries that represent a "decisive share of global production" in the sectors or subsectors deemed exposed to carbon leakage (1) "firmly" commit to reducing greenhouse gas emissions in those sectors or subsectors "to any extent comparable to that of the Community and within the same time frame," and (2) have installations located in their countries whose carbon efficiency is "comparable" to that of the Community. However, with respect to the second factor, the EC states:

> As regards the carbon efficiency, the relevant data necessary for that assessment is not available due to incomparability of statistical definitions and general lack of global data at the required level of disaggregation and sectoral detail. Therefore, the criteria set out in Article 1 0a(1 8) of Directive 2003/87/EC had no effect on the list of sectors and subsectors. (paragraph 22)

U.S Proposals to Address Carbon Leakage: H.R. 2454

Greenhouse gas reduction legislation introduced over the last two Congresses has included provisions to address carbon leakage. In general, two strategies have been employed: (1) free allocation of allowances (similar to that of the EU); and (2) an international reserve allowance (IRA) scheme. H.R. 2454, as passed by the House, contains both of these strategies.[34]

Title IV of H.R. 2454 would amend the bill's new Title VII of the Clean Air Act by creating a new Part F to address carbon leakage. The purpose of the new Part F is both environmental, in terms of reducing potential carbon leakage resulting from potential shifts of production and investment from the United States to countries without carbon controls, and economic, in terms of preventing the associated job loss from such a shift. Specifically, the purposes of Part F as a whole would be (1) "to promote a strong global effort to significantly reduce greenhouse gas emissions and, through this global effort, stabilize greenhouse gas concentrations in the atmosphere at a level that will prevent dangerous anthropogenic interference with the climate system," and (2) "to prevent an increase in greenhouse gas emissions in countries other than the United States as a result of direct and indirect compliance costs incurred under" the new Title VIII.[35]

The free allocation scheme (subpart 1) would be further aimed at the following: (1) "to provide a rebate to the owners of and operators of entities in domestic eligible industrial sectors for their greenhouse gas emissions costs incurred under this title, but not for costs associated with other related or unrelated market dynamics"; (2) "to design such rebates in a way that will prevent carbon leakage while also rewarding innovations and facility-level investments in energy efficiency performance improvements"; and (3) "to eliminate or reduce distribution of emission allowances under subpart 1 when such distribution is no longer necessary to prevent carbon leakage from eligible industrial sectors."[36]

The IRA scheme (subpart 2) would have these additional purposes: (1) "to induce foreign countries, and, in particular, fast-growing developing countries, to take substantial action with respect to their greenhouse gas emissions consistent with the Bali Action Plan developed under the United Nations Framework Convention on Climate Change" and (2) "to ensure that the measures described in subpart 2 are designed and implemented consistent with applicable international agreements to which the United States is a party."[37]

H.R. 2454, Title IV, Subpart 1: Free Allocation of Allowances

Description of Rebate Program

Subpart 1 of the new Part F would create a rebate program directed at energy/greenhouse gas- intensive, trade-exposed industries harmed by the direct emissions reduction costs and indirect increased energy input costs from implementing Title VII (the cap-and-trade provisions of H.R 2454). The program would begin by requiring EPA to publish a list of eligible industrial sectors and amount of allowances to be rebated per unit of production for the next two years by June 30, 2011 (revised every four years thereafter). Presumptively eligible industrial sectors would be determined at the six-digit classification level in Codes 31-33 of the North American Industrial Classification System of 2002 (NAICS).[38] As determined by EPA, presumptively eligible sectors, based on six-digit NAICS classification, are those that (1) meet energy or greenhouse gas intensity criteria (specifically, that energy or greenhouse gas costs are at least 5% of the value of their shipments) and trade exposure criteria (specifically, a trade intensity of at least 15%, based on the value of a sector's total imports and exports divided by the value of its shipments and imports); or (2) have very high energy or greenhouse gas intensity (at least 20%). The bill specifies data sources to be used in these determinations and, specifically, annual average data for 2004-2006, unless unavailable. However, the bill provides that EPA shall determine additional sectors to be eligible if they (1) meet the greenhouse gas or energy intensity criteria at the time the rule is promulgated and (2) meet trade intensity criteria based on post-2006 data. The bill also has provisions allowing individual entities to petition for inclusion of their sub sector under the program (Section 763).[39] Potential coverage is focused on primary products, such as iron, steel, aluminum, and cement. The bill expressly prohibits the petroleum refining sector from being considered an "eligible industrial sector."[40]

Based on the best data available, EPA is to provide the rebate to eligible companies based on a two-part formula: (1) 100% of the industry's annual average emissions per unit of output over the most recent four years multiplied by the company's annual average output over the preceding two years (direct emissions); and (2) average emissions per kilowatt-hour of electricity purchased by the company multiplied by the industry average electricity used per unit of output over the preceding two years multiplied by an electricity efficiency factor to be determined by EPA (indirect emissions). Entities not covered by Title VII are eligible for the indirect emissions rebate. If these formulas result in more allowance needs than provided under the bill, the allocations to entities would be reduced on a pro rata basis to match the allowances available (Section 764).

Unless modified by the President, the allowance rebates are phased out over a 10-year period, beginning in 2026. Facilities that ceased to engage in qualifying activities would lose their allocations at the point they ceased those activities. As provided in Sec. 767, the President may modify the phase-out schedule for a sector if 15% or more of U.S. imports for that sector is still produced in countries with inadequate carbon policies.

Eligible Industries

The designation of six-digit NAICS codes for determining eligibility adds a level of precision to the program that could make implementation more straightforward than would otherwise be the case. While there are about 450 manufacturing sectors designated at this level within these three codes,[41] it is likely that less than 50 of these would be deemed presumptively eligible under the detailed requirements set out in the bills. During deliberations on H.R. 2454, the Energy-Intensive Manufacturers' Working Group on Greenhouse Gas Regulation provided detailed testimony on the energy intensity and trade intensity of the U.S. manufacturing sector.[42] These data, based on analysis done for the Working Group by FTI Consulting, are presented in Figure 5 and Figure 6. According to the Working Group, 47 sectors are presumptively covered under subpart 1. This number is considerably less than the 164 sectors and subsectors potentially covered under the EC decision. The primary reason for the difference is that the main set of criteria incorporated in subpart 1 includes all three of the factors discussed earlier: (1) the greenhouse gas intensity of a company's products which influences the company's profitability and the products' cost; (2) the company's ability to pass on any increased costs to consumers without losing market share or profitability; and (3) the company's ability to mitigate carbon emissions, reducing the impact of the carbon policy on its operations and profitability.[43]

The EPA has also compiled a list of presumptively covered sectors; a list that is identical to that above, with two exceptions: paperboard mills (322130) and beet sugar (311313) are not included in the EPA list due to differences in data sources.[44]

Proposed Funding

Under H.R. 2454, energy-intensive, trade-exposed industries are allocated 2% of available allowances in 2012 and 2013, 15% of available allowances in 2014, and 13.4% of available allowances from 2015 through 2025. In addition, H.R. 2454 mandates that energy-intensive, trade-exposed industries receive their share of allowance value provided local electric distribution companies (LDCs) for electricity rebates. According to 2006 Bureau of Census data, the eligible industries purchased between 295 billion (EPA list) and 315 billion (Working Group list) kilowatt-hours of electricity.[45] They represent between 29% and 31% of retail sales to the industrial sector in 2006, or between 8% and 8.6% of total retail sales. Assuming a pass-through of allowance value by the LDCs based on 2006 data, this would represent about 2.4% to 2.6% of available allowances.

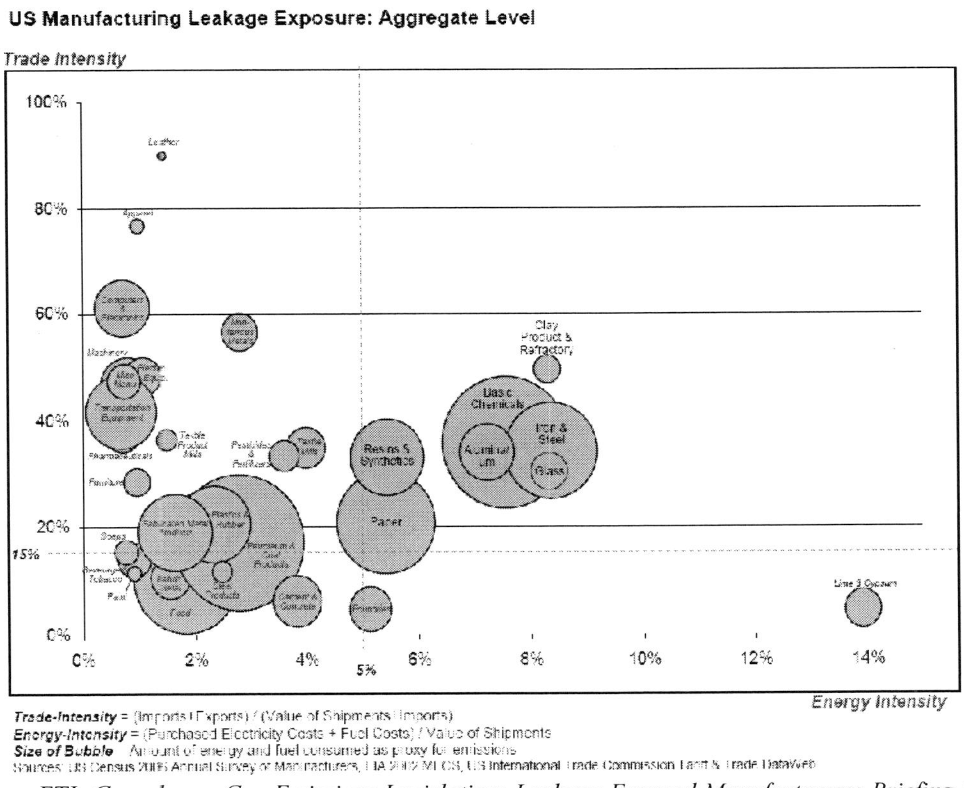

Source: FTI, *Greenhouse Gas Emissions Legislation: Leakage-Exposed Manufacturers: Briefing Book* (June 2009).

Figure 5. U.S. Manufacturing Exposed to Carbon Leakage Risk

Qualifying Manufacturing Industries

Sectors	NAICS	Sub-Industry	Leakage-Intensity Energy Intensity	Trade Intensity	Value of Shipments
Pulp, Paper & Newsprint Mills	322110	Pulp mills	8%	89%	4
	322121	Paper (except newsprint) mills	8	24	47
	322122	Newsprint mill products	8	64	4
	322130	Paperboard mills	12	22	23
Basic Chemicals	325110	Petrochemicals	12	17	66
	325131	Inorganic dyes and pigments	5	55	4
	325132	Synthetic organic dyes and pigments	5	39	3
	325181	Alkalies and chlorine	25	27	6
	325182	Carbon black	8	26	2
	325188	All other basic inorganic chemicals	8	56	19
	325191	Gum and wood chemicals	7	26	1
	325192	Cyclic crude and intermediates	7	76	9
	325193	Ethyl alcohol	7	10	8
	325199	All other basic organic chemicals	7	52	69
	325211	Plastics material and resins	5	36	79
	325212	Synthetic rubber	6	56	7
	325221	Cellulosic organic fibers	6	53	1
	325222	Non-cellulosic organic fibers	6	37	7
Nitrogenous fertilizer	325311	Nitrogenous fertilizer	12	81	4
Ceramics/Porcelain	327111	Vitreous china plumbing fixtures	5	49	1
	327112	Vitreous china and earthenware articles	5	85	1
	327113	Porcelain electrical supplies	5	29	1
	327122	Ceramic wall and floor tiles	7	68	1
	327123	Other structural clay products	10	27	0.2
	327124	Clay refractory	5	26	1
	327125	Non-clay refractory	5	41	1
Glass Production	327211	Flat glass	16	44	3
	327212	Other pressed and blown glass and glassware; incl. optical fiber	11	57	4
	327213	Glass containers	14	19	4
Cement	327310	Cement	15	18	11
Lime	327410	Lime	23	3	2
Fiberglass	327993	Mineral wool	8	17	6
Iron & Steel	331111	Iron and steel	7	34	92
	331112	Electrometallurgical ferroalloy products	7	71	1
	331210	Iron and steel pipe and tube from purchased steel			10
	212210	Iron ore mining and beneficiation	18	54	2
Alumina/um	331311	Alumina refining	21	65	1
	331312	Primary aluminum production	22	64	6
Copper	331411	Primary smelting and refining of copper	6	71	10
	212234	Copper and nickel mining and beneficiation			
Other Industries	311221	Wet corn milling	10	19	10
	311313	Beet sugar	6	17	3
	314992	Tire cord and tire fabric mills	5	32	1
	321219	Reconstituted wood products	7	38	8
	327992	Ground or treated minerals and earth	9	17	3
	331419	Primary nonferrous metal (except copper and aluminum)	7	69	5
	335991	Carbon and graphite products	6	49	2

Trade-Intensity = (Imports+Exports) / (Value of Shipments+Imports)
Energy-Intensity = (Energy & Fuel Costs + Generation) / Value of Shipments
Sources: US Census 2006 Annual Survey of Manufacturers, EIA 2002 MECS, US International Trade Commission Tariff & Trade DataWeb
Note: Percentages are rounded to nearest percentage point

Source: FTI, *Greenhouse Gas Emissions Legislation: Leakage-Exposed Manufacturers: Briefing Book* (June 2009)

Figure 6. Presumptively Eligible Energy-Intensive, Trade-Exposed Industries

The allowances these allocations represent are presented in **Figure 7**. After 2025, the allocation to energy-intensive, trade-exposed industries is phased-out over a 10-year period. The allocation to LDC is phased-out over a five-year period, beginning after 2025.

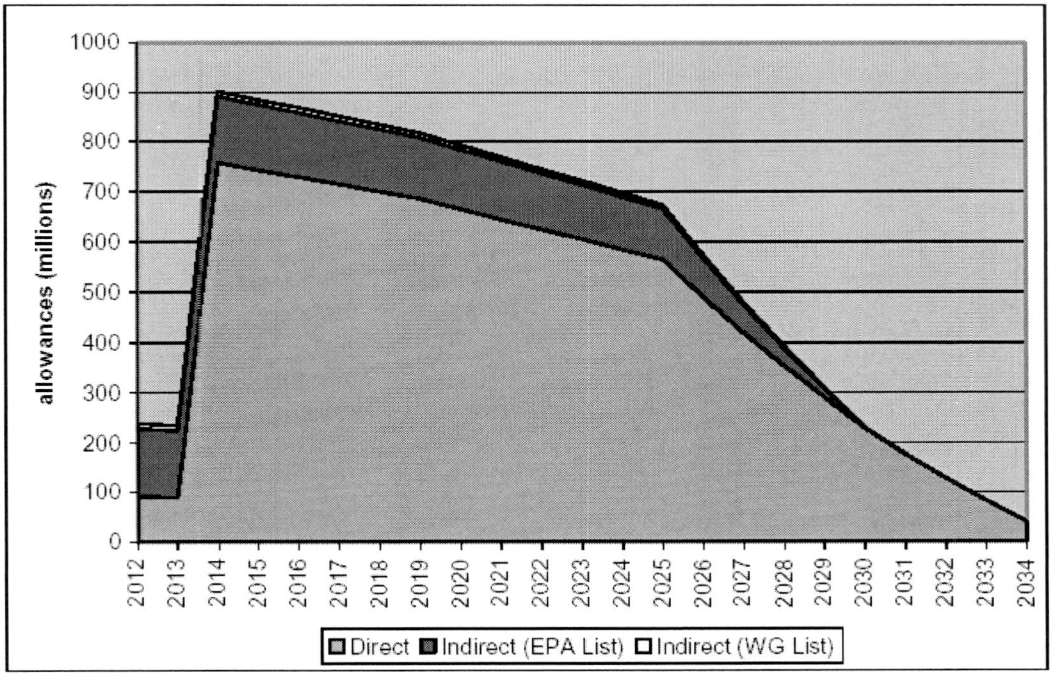

Source: H.R. 2454 and CRS calculations.
Notes: See text.

Figure 7. Direct and Indirect Allowance Allocations to Energy-Intensive, Trade-Exposed Industries under H.R. 2454

Analysis

Adequacy of Allocation

Both EPA and FTI (for the Working Group) have estimated the required allowances to compensate eligible industries for their direct and indirect greenhouse gas emissions. According to EPA, the total is 738 million allowances annually from 2014 through 2025, or 14.5% of available allowances; for the Working Group, the total is 828 million allowances. The differences result from higher estimates by the Working Group for paperboard, cement, plastics, iron and steel processes, and phosphates and soda ash. A breakdown of annual emissions by direct and indirect sources is provided in Table 1. In addition, the Working Group recommends an additional 10% be included as a reserve for individual showings of need and for methodological uncertainty. These contingencies raise the Working Group's estimate to 910 million allowances annually, or 16.2% of available allowances.

As noted above, after providing 15% of available allowances to energy-intensive, trade-exposed industries in 2014, H.R. 2454 provides 13.4% from 2015 through 2025. This would compensate between 82% and 92% of the industries projected direct and indirect costs. In addition, according to H.R. 2454, all industry is eligible for the pass-through of allowance value provided via local electric distribution companies (LDCs) (Sec. 783(b)(5)(D)). As noted above, based on Census Bureau and Energy Information Administration (EIA) data, CRS calculates a "ballpark" estimate of an additional 2.4% to 2.6% of available allowances being directed toward cost relief for energy-intensive, trade-exposed industries.[46] This estimate is considerably lower than the roughly 5% estimate provided to CRS by EPA. However, EPA notes that its numbers are overestimates of actual amounts because their model's (ADAGE) energy-intensive manufacturing sector is "much larger" than the industries made eligible by the language in H.R. 2454.[47]

Table 1. Direct and Indirect Emissions from Eligible Industries (annually, in million Metric Tons of CO_2 Equivalent)

Source	Direct Emissions from Combustion	Direct Emissions from Industrial Processes	Indirect Emissions from Electricity Consumption	Total
EPA	383	173	183	738
FTI	413	198	216	828

Source: U.S. Environmental Protection Agency, *Comparison of FTI and EPA analyses of H.R. 2454, Title IV*, Memorandum to the House Energy and Commerce Committee Staff (June 10, 2009).

Notes: EPA estimates are based on the average of 2004-2006 emissions, assuming no growth or efficiency improvements through 2025. FTI estimates are based on 2007 emissions, assuming no growth or efficiency improvements through 2005. Estimates do not include the reserve for individual showings of need and for methodological uncertainty included in the Working Group estimate of 910 million metric tons of CO_2 equivalent.

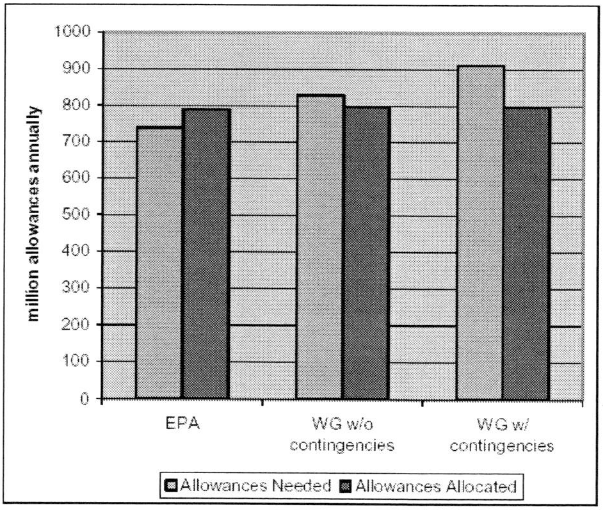

Source: CRS calculations and U.S. Environmental Protection Agency, *Comparison of FTI and EPA analyses of H.R. 2454, Title IV*, Memorandum to the House Energy and Commerce Committee Staff (June 10, 2009).

Notes: EPA estimates are based on the average of 2004-2006 emissions, assuming no growth or efficiency improvements through 2025. FTI estimates are based on 2007 emissions, assuming no growth or efficiency improvements through 2005. The Working Group w/o contingencies estimates do not include the reserve for individual showings of need and for methodological uncertainty included in the Working Group w/ contingencies estimate.

Figure 8. Projected Allowance Need and Allocation to Eligible Industries (average annual, 2014-2025)

As indicated by Figure 8, H.R. 2454's allocation scheme would appear to provide sufficient allowances if EPA's estimates are correct. If the Working Group's estimates are correct, or if individual showings of eligibility prove significant, the pool of allowances provided by the bill would not be adequate under the assumptions used here.

Whether the individual state public utility commissions (PUCs) (or other responsible body in the case of publicly owned utilities or cooperatives) would work to ensure that LDCs did return to industry the share provided it in the bill or would attempt to tilt allocations in favor of residential consumers is disputed. The language of H.R. 2454 is clear with respect to providing energy- intensive, trade-exposed industries with their share of the electricity rebate (Sec. 783(b)(5)), and each allowance misused by a state would be considered a separate violation of the Clean Air Act. H.R. 2454 requires a representative sample of LDCs to submit an annual report on implementation of the electricity rebate; this is to include how they are complying with the requirement to provide allowance value to energy-intensive, trade-exposed industries. A question is whether or not this reporting requirement, along with EPA implementation of the enforcement provision, would sufficiently influence LDC and PUC decision-making.

The Working Group has expressed great skepticism about the states' public utilities commissions' willingness to pass through savings to industry instead of favoring residential consumers—a decision over which Congress would have limited influence. This skepticism may not be unfounded. In an October 28, 2009, letter to Chairman Boxer and Ranking Member Inhofe of the Senate Environment and Public Works Committee, the National Association of Regulatory Utility Commissioners (NARUC) urged the Senate to think carefully before "handcuffing" state regulators. As stated in the letter:

> NARUC understands the need for federal oversight of what will undoubtedly be a significant amount of money flowing between LDCs and consumers. However, we also believe that State commissions are far more accountable to ratepayers than distant bureaucracies in Washington, and are far more efficient at developing innovative and entrepreneurial clean energy programs. State commissions know their localities and constituents best, and we are obligated to ensure fair, just and reasonable rates. The Senate should give States more leeway in distributing allowance proceeds so consumers can truly benefit.[48]

Effectiveness of Free Allowance Scheme

Both the EPA and the Energy Information Administration (EIA) have explicitly examined the impact of H.R. 2454's free allowance allocation to energy-intensive, trade-exposed industries. In the EPA/ADAGE analysis, energy-intensive manufacturing output is projected to decline by 0.3% from base case levels in 2015 and by 0.7% in 2020 without H.R. 2454's free allocation scheme. With the free allocation scheme, energy intensive manufacturing output is projected to increase by 0.04% from base case levels in 2015, and then decline by 0.3% from base case levels in 2020.[49] The free allocation scheme phases out in the 2020s.

The EIA/NEMS analysis of energy-intensive, trade-exposed industries also indicates that the free allocation to those industries reduces the impact of H.R. 2454 that they would otherwise bear. As stated by EIA:

> Receiving these permits ameliorates the impact of increased energy prices and therefore industries face energy prices that are not impacted by the permit values. As a result, when energy prices increase, the reductions in output of these trade- and energy-vulnerable industries are less than overall manufacturing impacts and mirror the impacts (in terms of percentage change from the Reference Case) of total industrial shipments. In past EIA analysis of industrial impacts of energy price increases, these energy-intensive industries typically experience larger losses compared to overall manufacturing.[50] [footnotes omitted]

The overall effect of the free allocation over time can been seen in Figure 9, from the EIA report. As indicated, the impact on energy-intensive, trade-exposed industries is comparable to that on industry as a whole, suggesting that the allowance allocation has a positive effect in alleviating any disadvantage they may have from being exposed to international competition from countries without comparable carbon policies.

Although the scheme would appear effective in mitigating the trade-related impact of the program on energy-intensive, trade-exposed industries, production cost for those industries (along with other industries) could increase because of the potential pass-through of compliance-related costs by upstream producers of various inputs into their manufacturing processes (e.g., feedstocks, petroleum, etc.). Whether these costs would become significant would depend on the ability of upstream suppliers to pass on those costs, and the ability of the downstream industries to respond by increasing the efficiency of their operations or by substituting other, less-costly inputs into their processes.

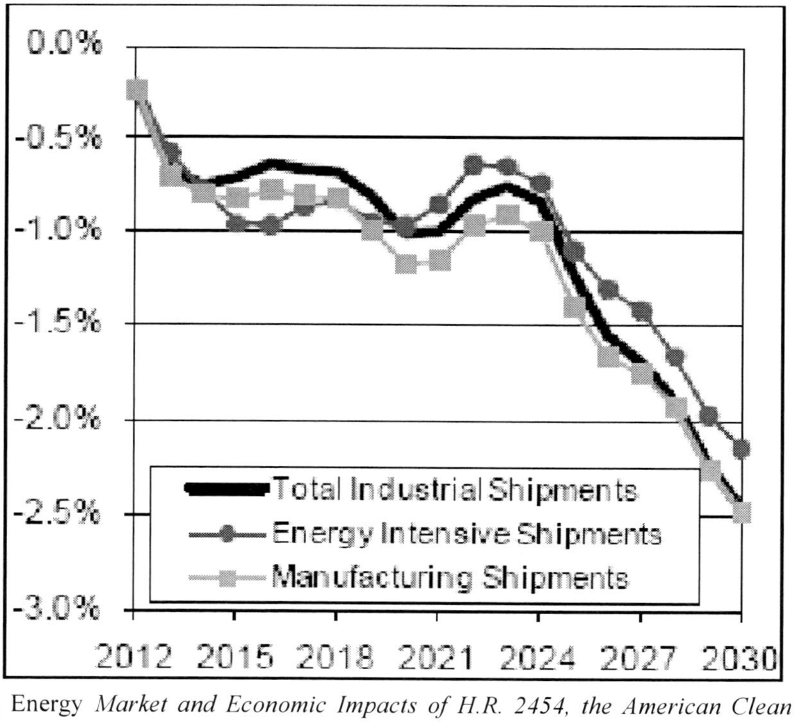

Source: EIA, Energy *Market and Economic Impacts of H.R. 2454, the American Clean Energy and Security Act of 2009* (August 2009), p. 45.

Figure 9. Industrial Impacts in the H.R. 2454 Basic Case, 2012-2030 (percent change from Reference Case)

Phase-Out Schedule

H.R. 2454 provides that, unless modified by the President, the allowance rebates are phased out over a 10-year period, beginning in 2026. In addition, the pass-through of allowance value from LDCs is also phased-out, but on a shorter schedule (beginning in 2026 and reaching zero in 2030) As provided in Section 767, the President may modify the phase-out schedule for the direct rebate for a sector if 15% or more of U.S. imports for that sector are still produced by countries with inadequate carbon policies.[51] There is no such authority for extending the pass-through received via the LDCs. As suggested by Figure 9, the EIA analysis assumes the phase-out begins on schedule in 2026 and the result is declining output for energy intensive, trade-exposed industries. This raises questions about the timing of any phase-out, and the extent to which the possible modification of the phase-out schedule introduces uncertainty in corporate decision-making.

H.R. 2454, Title IV, Subpart 2: International Reserve Allowance Scheme

Description of Program

Overview

If implemented, Title IV, subpart 2 of H.R. 2454 would require EPA to establish an international reserve allowance scheme that would essentially impose a shadow allowance requirement on importers of greenhouse gas-intensive, trade-exposed products, creating a de facto tariff. Basically, the scheme would require importers of energy-intensive products from countries with insufficient carbon policies to submit a prescribed amount of "international reserve allowances," or IRAs, for their products to gain entry into the United States. Based on the greenhouse gas emissions generated in the production process, IRAs would be submitted on a per-unit basis for each category of covered goods from a covered country. Specifically, Section 768 requires EPA to promulgate rules establishing an international reserve allowance system for covered goods from the eligible industrial sector, including allowance trading, banking, pricing, and submission requirements.

While subpart 1 would limit the distribution of emission allowances to eligible industrial sectors, Part F's definition of the term "covered goods," a term used only in subpart 2, goes beyond goods produced by eligible industrial sectors to include a "manufactured item for consumption" (i.e., finished goods, which could involve items ranging from aluminum cans to automobiles). [52] Allowances would potentially be required for importation into the United States of goods from a covered country that correspond to goods produced by U.S. eligible industrial sectors and, in some cases, for the importation of manufactured items for consumption from such countries.

The program would need to be consistent with U.S. commitments under international agreements, and in a manner that minimizes the likelihood of carbon leakage resulting from cost differentials resulting from compliance by U.S. companies with the U.S. reduction program compared with compliance by foreign companies with their nation's reduction program. The EPA would be required to adjust the international reserve allowance requirement based on the value of allowances allocated free under subpart 1 and under

Section 782(a) (electricity providers), including reducing the requirement to zero. The international reserve allowances issued under this program may not be used by covered entities to comply with the domestic emissions cap under Title VII. Also, this program may not apply to imports entering the United States before January 1, 2020.

Initial Action: Section 765

Under Section 765, the President is required as soon as practicable after enactment to notify all non-exempted countries that the United States (1) seeks international agreements that commit all major emitting nations to contribute equitably to reducing greenhouse gas emissions; (2) requests the country take appropriate measures to limit its greenhouse gas emissions; and (3) may apply the international reserve requirements of this subpart to a covered good beginning on January 1, 2020. Exemptions are provided under section 768(a)(1)(E) for the (1) least developed countries, (2) countries that emit less than 0.5% of global greenhouse gas emissions and have minimal export trade with the United States in covered sectoral products, and (3) countries meeting the comparability criteria of Section 767 (discussed below).

Section 766 states the environmental and economic elements the United States would seek in negotiating an international greenhouse gas reduction agreement.

Further Requirements and Criteria: Section 767

The President is further required by January 1, 2017 (and biannually thereafter), to submit a report to Congress on the effectiveness of the emission rebates under Subtitle 1 at mitigating carbon leakage and recommendations on improving the subtitle's purposes.[53] The report must also include an assessment, for each industrial sector receiving rebates, as to whether, and by how much, the per unit cost of production has increased for the sector, taking into account the provision of the rebates to the sector and the benefit received by the sector from the provision of free allowances to electricity providers under new section 7 82(a). In addition, the report must contain recommendations on improving the purposes of subpart 2, including an assessment of whether an IRA program for the eligible industrial sector would be feasible and useful. Further, to the extent that the President determines that an IRA program would not benefit a particular eligible industrial sector because its exposure to carbon leakage is due to competition in third country markets (i.e., occurs because of the sector's export trade), the President would need to identify alternative actions or programs consistent with the purposes of subpart 2. The President could also determine in such a case that an IRA program will not apply to the sector, though the determination must be approved by Congress (see below). Finally, the report must assess the amount and duration of assistance, including the distribution of free emission allowances, being provided to industrial sectors in other developed countries to mitigate compliance costs for domestic greenhouse gas (GHG)-reduction in those countries.

In addition, unless there is a multilateral agreement on reducing greenhouse gases in force for the United States by January 1, 2018, the President would be required to establish an international reserve allowance program for all eligible sectors unless the President determines, and the Congress concurs, that a sector covered under the program, or inclusion of a sector within that program, would not be in the nation's economic or environmental interests.[54] To become effective, each such determination would need to be approved by both houses of Congress within 90 days after the President submitted his determination.[55] Precisely when such a presidential determination and congressional concurrence must occur is not explicitly stated; however, a strict interpretation would suggest it must occur before the President is required to make his next determination under the bill's provisions on June 30, 2018.

Beginning June 30, 2018, and every four years thereafter, the President would be required to determine for each eligible industrial sector whether more than 85% of U.S. imports of "covered goods" for that sector are produced or manufactured in countries that meet one of these criteria:

> (1) the country is party to "an international agreement to which the United States is a party that includes a nationally enforceable and economy-wide greenhouse gas emissions reduction commitment for that country that is at least as stringent as that of the United States";
>
> (2) the country is a party to a multilateral or bilateral emission reduction agreement *for that sector* to which the United States is a party; or
>
> (3) the country has annual energy or GHG intensity for the sector comparable to or less than the energy or GHG intensity for the sector in the United States for the most recent year for which data are available.[56]

The bill does not appear to specify a time period within which the imports used in the calculation must have entered the United States, nor does it specify whether the quantity of imports is to be calculated on the basis of the value of the imports or on the basis of output (i.e., units imported).[57]

If the 85% threshold is not exceeded, the President would be required to assess the effectiveness of both rebates (including the benefit that the sector receives from the provision of free allowances to electricity providers) and an IRA program in addressing or mitigating, or potentially addressing or mitigating, carbon leakage in that sector. The President would then need to respond by (1) modifying the rebate formula *and* (2) implementing (or, in the case of future determinations, continuing to implement) an IRA program for the sector.[58] If the threshold is exceeded, however, the President would be expressly prohibited from applying a sectoral IRA program.[59]

Effectively, the international reserve allowance program would be established for all eligible sectors unless the Congress (or the Senate, in the case of a treaty) approves a multilateral agreement reducing greenhouse gases and the agreement enters into force for the United States, or the Congress votes to concur with a presidential determination that including an eligible sector would not be in the nation's economic or environmental interest. Further, once the program is established for a sector, H.R. 2454 would not permit the President to determine, as a result of his assessments, whether or not the rebate formula should be altered or an IRA program should be applied. By not providing for this intermediate step, the bill would effectively make these two actions mandatory once the President had determined that the 85% threshold had not been exceeded for the sector involved. In the event a program is to be applied, the bill would prohibit IRAs from being collected on goods imported into the United States before January 1, 2020.[60]

EPA Implementing Regulations: Section 768

If implemented, section 768 requires that the regulations that EPA issues for an IRA program for an eligible industrial sector contain specific elements. The regulations must be issued with the concurrence of U.S. Customs and Border Protection (CBP), which has general statutory responsibility over the entry of goods into the United States, including the assessment and collection of duties and fees on imported products. Such regulations must

- establish an IRA program for the sale, exchange, purchase, transfer, and banking of IRAs for covered goods with respect to the sector;
- ensure that the price for purchasing IRAs from the United States on a particular day is equivalent to the auction clearing price for emissions allowances [under the new cap-andtrade provisions of Title VII] for the most recent emission allowance auction;
- establish a general methodology for calculating the quantity of IRAs that a U.S. importer of any covered good must submit;
- require the submission of appropriate amounts of IRAs for covered goods with respect to the eligible industrial sector that enter U.S. customs territory;
- specify the procedures that CBP will apply for the declaration and entry of the sector's covered goods into U.S. customs territory;
- establish procedures that prevent circumvention of the IRA requirement for covered goods that are manufactured or processed in more than one foreign country. [61]

In establishing a general methodology for calculating the required number of IRAs for a covered good, EPA would be required to include an adjustment based on the value of rebates distributed to the eligible industrial sector involved as well as the benefit received by the sector from free allowances received by electricity providers under the bill.[62] In applying such an adjustment, EPA would be permitted to determine that the amount of IRAs for a given product should be zero.

In addition, in administering a sectoral IRA program, EPA would need to exempt goods originating in three categories of countries from the border IRA requirement: (1) countries meeting any of the three standards set out in the new Section 767 for determining if a country had taken adequate action to reduce its GHG emissions for a sector, and ultimately, whether the corresponding U.S. industrial sector merited further assistance (see below); (2) the least developed of developing countries (LDDCs); and (3) countries that the United States determines are *de minimis* emitters responsible for less than 0.5% of total global greenhouse gas emissions and for less than 5% of U.S. imports of covered goods for an eligible industrial sector.[63]

The bill would also require the EPA to establish the IRA program "consistent with international agreements to which the United States is a party."[64] Absent a limiting definition in the bills, this requirement would seemingly encompass all U.S. international agreements, including both environmental agreements and international trade agreements.

Decision-Making Process

Figure 10 is a flow chart that traces the decision-making process of the International Reserve Allowance scheme created by the Center for Clean Air Policy. [65]

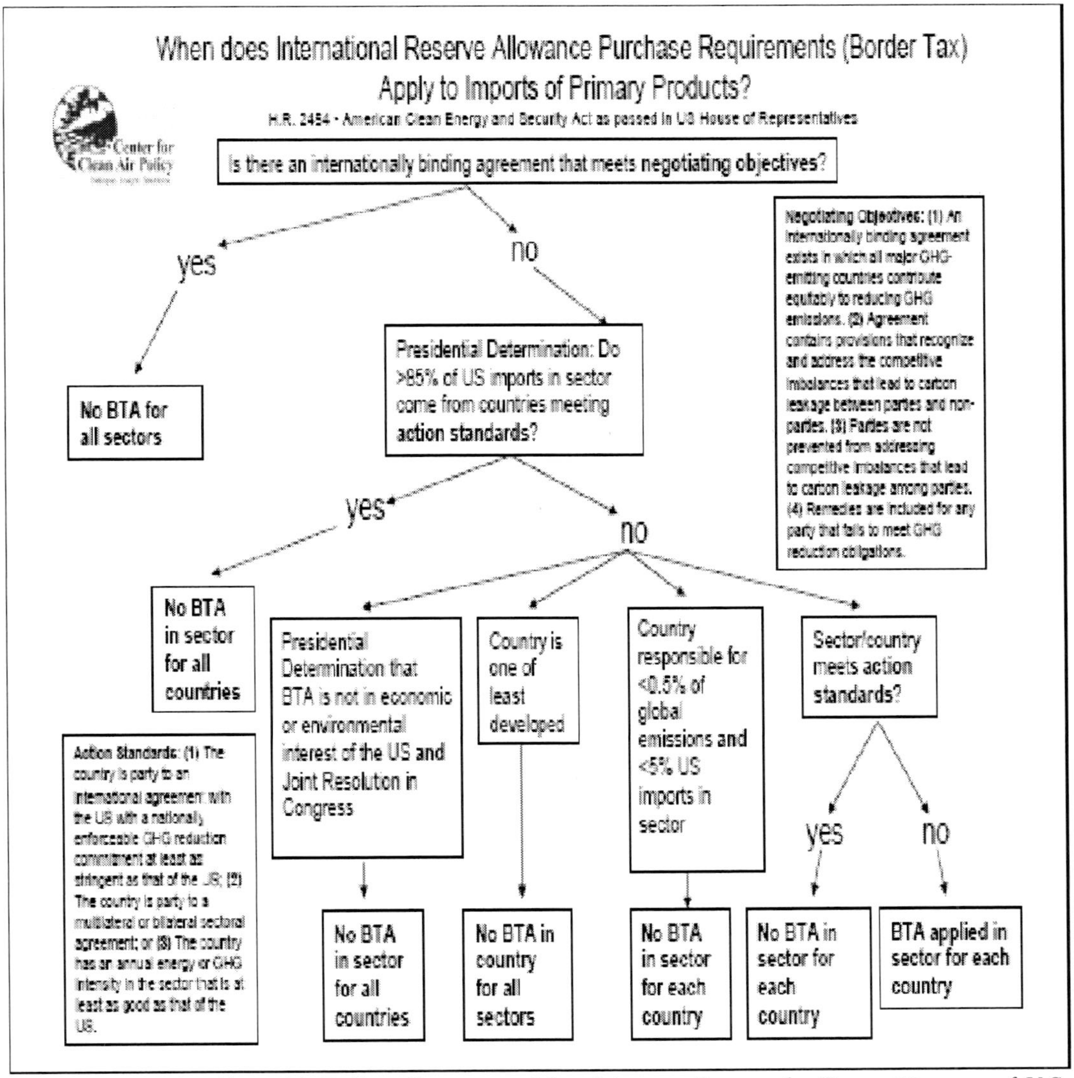

Source: Center for Clean Air Policy, *Summary of Provisions to Protect the Competitiveness of U.S. Industry in the American Clean Energy and Security Act* (July 23, 2009), p. 24.

Figure 10. Decision Tree for IRA Scheme

Analysis

Potential Impact

No analysis of subpart 2 and its impact on trade has been conducted. Indeed, the only analysis of an IRA scheme that has been done at all is one conducted by EPA (ADAGE) with respect to Title VI of S. 2191, introduced in the 110th Congress.[66] In that report, EPA's sensitivity analysis indicated that if countries without legally binding commitments to reduce greenhouse gases commit to maintaining their 2015 levels beginning in the year 2025, and to returning their emissions to 2000 levels by 2050, no international emission leakage occurred.

Imports of energy- intensive goods were projected to fall under this scenario, while exports expanded as developing countries coped with their new emission limits.

In a worst case scenario, EPA's 2008 sensitivity analysis looked at a no-international-actions-to2050 scenario. In this scenario, the International Reserve Allowance provisions of S. 2191 were assumed to be triggered because of the lack of international action. Emissions from countries without legally binding commitments were estimated to rise by 350 million metric tons of CO2e by 2030 and 385 million metric tons by 2050—less than 1% of their base case levels under ADAGE. It would have been equivalent to U.S. emission leakage rates of approximately 11% in 2030 and 8% in 2050. These emissions compared with increases of 361 million metric tons and 412 million metric tons for 2030 and 2050, respectively, if the IRA provisions were not implemented. EPA described the impact of the IRA program on leakage as "minimal."[67]

The projected impact on imports was more significant. Without the International Reserve Allowance requirement, imports from countries without legally binding commitments were projected to increase 5.4% in 2030, rising to 7% in 2050. In contrast, under the IRA provisions, imports were estimated to increase about 1% in 2030 and decline about 5% in 2050. U.S. exports declined in both cases as countries used more of their domestic manufacturing capacity. [68]

If the EPA projections for S. 2191 are transferable to H.R. 2454, the differential effect of IRA provisions on trade versus emissions leakage could present problems if the scheme is brought before the World Trade Organization (WTO).

In addition, this analysis does not fully account for the nature of international trade. Trade and economics involve dynamic processes that can respond to public policy in unanticipated ways. For example, trade sanctions based on primary goods, such as steel and aluminum, could have impacts on domestic downstream industries. An increase in the cost of raw steel or aluminum could drive up the costs of domestically manufactured finished products, such as automobiles, and encourage foreign countries to export more finished products to the United States. Indeed, a country could redirect its exports from primary goods to finished goods to avoid the trade sanctions. For example, South Korea, which exports both raw steel and automobiles, could focus its industrial policy toward automobile exports and away from raw steel exports. Thus, downstream companies that use greenhouse gas-intensive goods could have their competitiveness undermined by attempts to protect greenhouse gas-intensive, trade-exposed industries, particularly if their goods do not meet the criteria for "items manufactured for consumption" provided in the bill.

Data Needs

As noted earlier under the discussion of the EC's list of eligibility, lack of data prevented the EC from determining the carbon efficiency of installations in foreign countries (and thus their comparability with installations within the EU) as part of the criteria set out in Article 1 0a(1 8) of Directive 2003/87/EC. Instead, the EC chose to ignore the issue. This is not an option under subpart 2.

While official emission data in the United States generally take about one to two years to be collected, quality assured, and published, many other countries do not have the infrastructure to create emission data on a timely basis. Under the United Nations Framework Convention on Climate Change (UNFCCC), the "most recent year" for emissions data for many countries is 1994. While U.S. emissions data are more reliable, the lack of equally reliable data for foreign countries may well prevent the United States from adequately determining whether the same conditions prevail in foreign countries and the United States (i.e., whether foreign GHG-reduction programs are in fact comparable to the United States). Moreover, the lack of reliable data may prevent the executive branch from properly determining whether the same conditions prevail in foreign countries relative to each other, that is, determinations may be made on different quantitative bases for different countries depending on data availability. The quality of data would also be a factor in determining which countries are not high emitters and are thus excluded from the import requirement altogether. Unreliable data may be particularly troublesome in implementing a statutory cutoff point and countries may fall just above or below the threshold.

For example, China submitted its "Initial National Communication on Climate Change" to the UNFCCC in October 2004.[69] The emission inventory included in that submission was for 1994. While China notes that its 1994 inventory was prepared in accordance with approved guidelines, uncertainties remained. It provides two reasons for the uncertainties:

> Firstly, as a developing country, China has a relatively weak position with regard to data, and in particular has many difficulties in obtaining activity data for estimating GHG emissions; Secondly, though sample surveys and on-the-spot examinations were carried out to some extent in the energy, industrial processes, agriculture, land-use change and forestry, and waste treatment sectors to collect the basic data for inventory development, the time span and specific sample observation points may not be fully representative due to the constraints in funding, time available and other factors.[70]

China is not alone. Emission data troubles exist for most "non-Annex 1" countries, that is, countries that are not subject to legally binding emission reductions under the Kyoto Protocol to the UNFCCC. As stated by the UNFCCC in its 2005 synthesis of initial national communications from non-Annex 1 countries: "Most Parties [non-Annex 1 countries] reported difficulties in preparing their GHG inventories, and indicated that their technical and institutional capacities were inadequate to meet their reporting obligations under the Convention for both the preparation and updating of national GHG inventories."[71]

Other data-related aspects of subpart 2 may also raise implementation issues. For example, as noted above, the most recent year for which official data on Chinese greenhouse gas emissions is available is 1994. In contrast, the most recent calendar year for which official data of Chinese production of major industrial goods (e.g., steel, iron, cement, etc.) are available is 2005.[72] This calculation may produce not only an uneven result for a particular country, but the results for different countries may vary depending on the years for which data are available and thus provide uneven results between countries. For example, in contrast to China's 1994 emissions inventory, the latest submission to the UNFCCC by South Korea provides an emissions inventory for 2001.[73] Multiply these differences between countries across affected sectors, subsectors, primary goods, and "items manufactured for consumption," and EPA's ability to create the matrix of data necessary to implement the scheme becomes problematic, at best.

Presidential Determination to Exclude a Sector

In the absence of a qualifying multilateral GHG-reduction agreement, H.R. 2454 would permit the President to determine that an IRA program should not be established for an eligible industrial sector because it would not be in the national economic or environmental interest of the United States to do so. As evident from possible contents of the President's initial report to Congress, legislators would have contemplated that such determinations might involve eligible export- dependent industrial sectors. The bill would not appear, however, to expressly preclude the President from making such a determination regarding any sector that he saw fit to exclude. Nevertheless, since Congress would have apparently distinguished between sectors depending on whether their primary trade exposure from the U.S. cap-and-trade program is on the import or export side, it may be more difficult to secure congressional approval for excluding a sector whose primary concern is the adverse effect of imports into the United States from countries that do not have GHG-reduction programs or have not made commitments to create them. This outcome may be even more likely given that there does not appear to be authority in the bill for the President to establish an IRA program for a sector once Congress has approved its exclusion. Whether Congress would agree with a presidential request to exclude a sector could be a key question for other countries engaged in negotiations on multilateral GHG-reduction agreements.

"Covered Goods"

An IRA program would apply to "covered goods" for an eligible industrial sector, defined in the legislation as any good, as identified by EPA, that is produced by the relevant sector, as well as any "manufactured item for consumption," that is, a good that "includes in substantial amounts one or more goods like the goods produced by an eligible industrial sector" (i.e., a downstream item such as car or refrigerator in the case of steel products).[74] Further, the bill appears to intend that an IRA program be in effect for that eligible industrial sector and that the product or products included in the downstream item be subject to IRA allowances greater than zero.[75] In addition, the industrial sector producing the good must have a trade intensity level of 15% or more and the producers of the good must demonstrate, and EPA must determine, that applying IRAs to the good is technically and administratively feasible and appropriate to achieve the purposes of Part F, taking into account the energy and GHG intensity of the sector producing the good (as determined under the formula that would be used to determine these levels for purposes of its qualifying as an "eligible industrial sector"), the ability of these producers to pass on cost increases, and "other appropriate factors."

While the bill would require that the sector producing the good meet the same trade intensity level needed to qualify as an eligible industrial sector, the bill appears to treat energy or GHG intensity differently, making the producers' existing levels, whatever they may be, a factor for the EPA to consider in deciding whether IRAs are feasible and appropriate to apply. In other words, in order to have their product be included as a "covered good," petitioning producers may not necessarily be expected to meet all the requirements for being deemed an "eligible industrial sector" for purposes of Part F, whether presumptively or by petition.

"Manufactured Items for Consumption" (Downstream Items)

By including manufactured items for consumption, H.R. 2454 would allow a good that otherwise would not be a product of a sector eligible for an IRA to be treated as if it were and the good would thus be a covered good that might qualify for the IRA program. As explained earlier, we are assuming that the legislation intends that an IRA program already be in effect for the eligible industrial sector producing the "like" input or inputs into the manufactured item and that the IRA requirement for the input or inputs is greater than zero. It is unclear, however, what the word "like" means for purposes of this threshold requirement. While it could, but does not necessarily, mean identical, how dissimilar the imported input or inputs could be from the goods that are produced domestically is not specified.[76]

Whether such a manufactured good would qualify for inclusion would depend in part on the trade intensity of the sector that produces the good. In addition, producers would seemingly need to provide EPcA with a strong factual and analytical basis to allow it to determine that applying IRAs to the product would be technically and administratively feasible and appropriate in the circumstances. As noted earlier, however, producers would not appear to be required to meet the energy and GHG intensity standards needed to qualify as an "eligible industrial sector" in making their case and, moreover, "other appropriate factors," unidentified in the bill, could enter into their argument and EPA's ultimate determination. Overall, while the rationale for including goods produced by an eligible industrial sector in an IRA program is clear, the application of an IRA program to manufactured goods for consumption could be problematic in that it would extend the benefits of an IRA program to an industry that would not have initially qualified as an eligible industrial sector and may yet have difficulty doing so.

Emphasis on International Action

As noted above, subpart 2 states that its purpose would be best achieved through international agreements negotiated by the United States and foreign countries and, to this end, states that it is U.S. policy to "work proactively under the United Nations Framework Convention on Climate Change, and in other appropriate fora, to establish binding agreements, including sectoral agreements, committing all major greenhouse gas-emitting nations to contribute equitably to the reduction of global greenhouse gas emissions."[77] The bill also sets out U.S. negotiating objectives for the multilateral environmental negotiations contemplated in the bill. These are

> to reach an "internationally binding" agreement in which all major GHG-emitting countries "contribute equitably" to the reduction of global GHG emissions;
>
> to include provisions "that recognize and address the competitive imbalances that lead to carbon leakage and may be created between parties and non-parties to the agreement in domestic and export markets" and not to prevent agreement parties from addressing "the competitive imbalances that lead to carbon leakage and may be created by the agreement among parties to the agreement in domestic and export markets"; and
>
> to include "agreed remedies" for any agreement party that fails to meet its GHG reduction obligations under the agreement.[78]

The bill also states that nothing in the negotiating objective involving competitive imbalances may be construed to require the United States to alter provisions of new section 764, providing for the distribution of emission allowance rebates.[79]

As discussed below, these objectives would be taken into account by the Senate or the Congress when it considers whether to approve any resulting multilateral GHG-reduction agreement.[80] Whether the United States is a party to such an agreement by January 1, 2018, would determine whether the President must initially establish IRA programs for eligible industrial sectors. Further, if such programs are established, the existence of a multilateral GHG-reduction agreement would be a factor used by the President in determining whether an IRA program should be applied with respect to a particular eligible sector for a given four-year period.

As noted above, the President would be required to establish an IRA program for each eligible industrial sector if, by January 1, 2018, a multilateral GHG-reduction agreement consistent with the bill's negotiating objectives has not entered into force for the United States, unless Congress has approved a presidential determination to exclude a sector.[81] H.R. 2454 would utilize the legislative approval process for the contemplated international agreement as a vehicle for Congress to indicate whether or not the agreement meets the legislative negotiating objectives outlined above. Thus, if, in submitting an agreement to the Senate or the Congress, the executive branch indicates that the agreement is consistent with these objectives, the agreement will be considered to be consistent as of the date that the Senate consents to the agreement or "legislation is enacted implementing such other agreement." [82] The Senate or the Congress may state in any such ratification or implementing measure, however, that the agreement should not be treated as consistent with these objectives for purposes of the requirement to establish IRA programs and for purposes of new section 768. It is under this section that EPA would issue regulations implementing an IRA program for each eligible industrial sector in the event that a qualifying agreement has not entered into force by January 1, 2018. As discussed below, any such regulations must exclude goods originating in any country that is a party to an international agreement to which the United States is also a party requiring a binding national GHG-reduction commitment as stringent as that of the United States.

Even if the multilateral GHG reduction agreement, by virtue of its approval by the Senate as a treaty or by the Congress as a congressional-executive agreement, were deemed to be consistent with legislative negotiating objectives as of a date that meets the January 1, 2018, deadline, the agreement would still need to meet the other requirement of new section 767(b)(1), namely, that the agreement has entered into force for the United States. For this to occur, the agreement itself would have to have entered into force, a situation that ordinarily occurs when an earlier agreed upon number of countries have accepted or acceded to it, and the United States would need to have deposited its instruments of ratification or accession with the entity designated under the agreement to receive them (i.e., officially accept the treaty or agreement obligations as a matter of international law and thereby become a party to it). This process raises the question of implementing legislation since the United States might not accede to a treaty or international agreement until any legislation needed to enable it to fully perform its treaty or agreement obligations under domestic law is enacted.[83] Thus, even though the bill, with its cap-and-trade program and other GHG-reduction provisions, would have been enacted into law to arrive at this point, further legislative action to implement the treaty or agreement could still conceivably be needed, a requirement that may further delay its entry into force for the United States.[84]

Some of these concerns may also arise with respect to the multilateral and bilateral sectoral agreements whose existence would be a factor in the President's determination as to whether the 85% import threshold is exceeded for an eligible industrial sector. The United States must be a party to any such agreement, and thus questions related to approval and implementation may need to be addressed. At the same time, H.R. 2454 is more lenient with respect to the elements of sectoral agreements than it is with respect to the multilateral GHG-reduction agreement that would moot the requirement that IRA programs be established or be taken into account in determining whether the import threshold was met. H.R. 2454 places no requirements on the content of a sectoral agreement, and thus the executive branch would seemingly have discretion to agree to sectoral GHG-reduction commitments that are weaker than some would like. In such case, increasing the percentage to be applied to the base figure (i.e., U.S. imports) in House- passed H.R. 2454 may have been a way of making it more difficult to reach the threshold when goods imported from one or more countries that are party to such sectoral agreements would be included within the calculation.

Table 2. Comparison of Top-20 Greenhouse Gas Emitting Countries (2005 data)

2005 Rank	Country	Annex 1	2005 GHG Emissions MMTCE	2005 Per Capita GHG Emissions (tons C/person)
1	China	No	1,970	1.5
2	United States	Yes	1,901	6.4
[3]	*European Union-27*	*Yes*[a]	*1,378*	*2.8*
3	Russian Federation	Yes	535	3.7
4	India	No	506	0.5
5	Japan	Yes	366	2.9
6	Brazil	No	277	1.5
7	Germany	Yes	267	3.2
8	Canada	Yes	200	6.2
9	United Kingdom	Yes	175	2.9
10	Mexico	No	172	1.7
11	Indonesia	No	162	0.7
12	Iran	No	155	2.2
13	Italy	Yes	154	2.6
14	France	Yes	150	2.5
15	Korea (South)	No	150	3.1
16	Australia	Yes	150	7.3
17	Ukraine	Yes	132	2.8
18	Spain	Yes	120	2.8
19	South Africa	No	115	2.5
20	Turkey	Yes	107	1.5
Total[b]			7,764	
	WORLD		10,569	1.6

Source: Climate Analysis Indicators Tool (CAIT) Version 6.0. (Washington, DC: World Resources Institute, 2008).

a. The Kyoto Agreement gave explicit authority to the original 15-member European Union to meet its obligations collectively; the EU has, in effect, expanded that authority as it has incorporated new members. If the EU-27 were ranked in terms of its 2005 GHG emissions, it would place 3rd.

b. Totals are of the 20 individual nations; they do not include the European Union.

Reactions from Other Countries: Defining "Comparable" Actions

There is a high probability of unintended consequences from subpart 2 as other countries react to the threat of a tariff. One potential consequence of subpart 2 is that foreign countries with more stringent carbon polices than those proposed in the United States could turn the tables and impose their own tariffs on U.S. goods exported to them. As discussed earlier, the EU has already agreed to a more stringent reduction program to the year 2020 than H.R. 2454 entails. Even if subpart 2 programs did not target the EU (because of the "comparable" provisions), it is conceivable that the EU might target the United States because of the U.S. lack of a reduction target "comparable" to that of the EU.

The argument about "comparability" could also extend to developing countries who are targeted by subpart 2. Targeted foreign countries could take the subpart's concept of comparability and employ a different metric—a metric more favorable to their situation—than the standard that subpart 2 would impose. For example, as illustrated in Table 2, developing countries could attempt to define comparability in terms of per capita greenhouse gas emissions. By that metric, China's greenhouse gas emissions are only one-quarter of those of the United States. For India, the metric is even more favorable; its emissions are only 8% of those of the United States. Based on this, or some other favorable metric, developing countries, such as China or India, could also turn the tables on the United States and impose their own tariffs on U.S. goods.

IMPLICATIONS FOR INTERNATIONAL TRADE OBLIGATIONS[85]

Both vehicles in Title IV of H.R. 2454 aimed at competitiveness and leakage concerns—the granting of free emission allowances to domestic firms and the imposition of border measures on imported products—raise issues involving U.S. obligations under World Trade Organization (WTO) agreements.[86] Legislation providing free emission allowances to carbon/energy-intensive trade-exposed industries may arguably confer a subsidy for purposes of the WTO Agreement on Subsidies and Countervailing Measures. Although the bill would require EPA to establish a border measure (IRA) program consistent with U.S. international agreements, a category that would include U.S. trade agreements,[87] a requirement that importers purchase IRAs to accompany particular imports might nonetheless be found to constitute a prohibited import surcharge or, if the product may not otherwise enter the United States, a prohibited quantitative restriction under the General Agreement on Tariffs and Trade 1994 (GATT).[88] If so, the requirement would need to be justified under a GATT exception to survive a WTO challenge. It is important to emphasize that while earlier GATT and WTO cases may provide a guide to the types of issues that may concern a WTO panel applying and interpreting GATT exceptions, measures are judged on a case-by-case basis. Thus, earlier decisions may not be fully predictive where a Member seeks to justify a novel and complex measure that affects a broad range of imported products, production processes, sources of manufacture, and trading partners.

Since the negotiating objectives set out in H.R. 2454 contemplate that a multilateral GHGreduction agreement may include provisions permitting, or at least not prohibiting, individual parties to address trade-related "competitive imbalances that lead to carbon leakage," it is possible that such an agreement could establish a set of principles or rights and obligations among the parties that address the allocation of emission allowances by WTO Member countries in the context of WTO subsidy obligations, provide scope for Members to impose border measures, or both.[89] A provision limiting the initiation of disputes for a defined period, a so-called "peace clause," might also be included.[90] Such an agreement might also be negotiated separate from multilateral climate change negotiations. WTO Members could obtain a WTO waiver for the provisions of such an agreement, incorporate binding commitments into WTO law by amending the relevant WTO agreements, or adopt separately negotiated principles or guidelines in a WTO decision.[91] Agreement on a binding WTO-related climate change accord is far from certain, however,[92] and thus, absent such an agreement or broad adherence thereto by WTO Members,[93] the WTO dispute settlement process may ultimately serve as the main forum for resolving WTO legal issues involving problematic trade-related climate change measures.

Disputes arising under WTO agreements are heard under the terms of the Understanding on Rules and Procedures Governing the Settlement of Disputes (Dispute Settlement Understanding or DSU).[94] Other WTO agreements, such as the WTO Agreement on Subsidies and Countervailing Measures, while providing for dispute settlement under the DSU rules and procedures, contain certain special and additional rules, which prevail over those in the DSU in the event of differences between the two. Dispute settlement is administered by the WTO Dispute Settlement Body (DSB), consisting of all WTO Members.

WTO dispute settlement may be characterized as a three-stage process, consisting of (1) consultations; (2) panel and possibly Appellate Body proceedings; and, (3) if a WTO decision is adverse to the defending Member, implementation. Once the DSB adopts panel and any Appellate Body reports finding that the defending Member has violated a WTO obligation, the defending Member would ordinarily be expected to withdraw the violative measure. [95] If the Member could not comply immediately, it would be given a reasonable period of time to do so. In the event that the defending Member fails to comply by the end of the established compliance period, the complaining Member may seek compensation from the defending Member or request authorization from the WTO to impose countermeasures (i.e., to suspend WTO concessions or other obligations owed the defending Member, usually, to place additional tariffs on selected products imported from the Member).[96] The DSU treats countermeasures as measures of "last resort," however, and permits them to be applied only as long as the measure found to violate WTO obligations remains in place or until the disputing parties settle their dispute in a mutually satisfactory way.

Certain actions by the DSB—namely, establishing a dispute settlement panel, adopting panel and Appellate Body reports, and authorizing a WTO Member to impose countermeasures—are virtually automatic; that is, the action will be taken unless all Members present at the DSB meeting agree not to do so ("reverse consensus" rule). The DSU contains an aspirational timeline of 18 months from the date a panel is established to the date a compliance period is determined. Complex cases are likely to require additional time, however, particularly at the panel stage.

Dispute settlement is generally Member-driven, so that it is up to the parties to a dispute to decide whether or not to take particular actions available to them (e.g., to request a panel, to request authorization to take countermeasures against a non-complying Member, or to apply such measures even if the WTO has authorized them). While the possibility of paying compensation or suffering the effects of retaliatory action may exert a degree of pressure on defending Members to comply with WTO decisions, and while DSU provisions indicate an overall intent that Members comply, the inclusion of compensation or retaliation as remedies, albeit temporary ones, recognizes that WTO Members may not always do so. In practice, Members have managed disputes at the implementation stage in a variety of ways short of taking retaliatory action.

Under GATT and now WTO dispute settlement practice, a WTO Member may challenge a measure of another Member "as such," "as applied," or both.[97] An "as such" claim challenges the measure as violative of a WTO agreement independent of its application in a specific situation and, as described by the WTO Appellate Body, seeks to prevent the defending Member from engaging in identified conduct before the fact.[98] Panels in past "as such" challenges have used an analytical tool known as the "mandatory/discretionary distinction," under which a law or regulation was considered not to violate a GATT or WTO obligation if it did not mandate a WTOinconsistent outcome or, in other words, could be applied in a WTO-consistent fashion.[99] If found to be discretionary under this analysis, the measure would need to be challenged "as applied." At the same time, the WTO Appellate Body, without examining the role of the distinction in a comprehensive way, has stated that the distinction should not be applied in a "mechanistic fashion,"[100] and thus the existence of discretionary elements in a statute or regulation may not necessarily shield it from an "as such" challenge.

Distribution of Free Emission Allowances

General Characteristics of Emission Allowances

An emission allowance may be defined as governmental permission to emit one ton of carbon dioxide or carbon dioxide-equivalent.[101] In a cap and trade system, recipients of emission allowances would include entities subject to emission caps and possibly other industrial entities that emit GHG gases directly in production processes and indirectly through the use of carbon- intensive fuels, as well as a broader array of entities that may be adversely affected by higher fuel prices resulting from compliance costs borne by capped fuel producers. The government may allocate allowances free of charge, require that they be obtained through an auction, or operate a mixed system incorporating both approaches.

Depending on its individual situation, a capped entity would use all of its allowances to cover emissions up to its annual cap; purchase additional allowances if it exceeded its cap and did not hold sufficient allowances to account for these excess emissions; or, in the event its annual emissions fell below the cap, sell unused allowances to other capped entities that need allowances to cover emissions that exceed their cap or bank them for future use or sale. Non-capped entities would either sell their allowances to capped entities or trade them in carbon markets. Thus, the situation of the recipient may differ depending on whether it is a capped or non-capped entity, and, if capped, whether the original allocation of allowances is sufficient, insufficient, or overgenerous.

Emission allowances have been recently characterized by the Congressional Budget Office (CBO) as "'cash-like' in nature" because they may be traded "in a large and liquid secondary market."[102] Moreover, in assessing the budgetary treatment of distributed allowances in a system where the government determines the scope of covered emissions and the number of allowances to be issued, CBO concluded that:

> the distribution by the federal government would be essentially equivalent to the distribution of cash grants, so CBO believes that such distributions should be treated as outlays. At the same time, allowances in a cap-and-trade system would be valuable financial instruments, so CBO thinks that the creation of allowances by the federal government should be recorded as revenues.
>
> That logic does not hinge on whether the government sells or, instead, gives away the allowances. Allowances would have significant value even if given away because the recipients could sell them, or if they are carbon dioxide emitters, use them to avoid incurring the cost of purchasing allowances or investing in costly emission mitigation mechanisms. Therefore, selling allowances and giving entities cash, and giving entities the allowances themselves and letting the entities realize their value, are essentially the same transaction. Sound budgeting requires that the budget treat equivalent transactions in the same way.[103]

In explaining its approach, CBO considers that the government grant of an allowance to a firm, business, or other recipient that would sell the allowance to a capped entity is a transaction that is "equivalent" to the government's taxing the capped firm or selling it an allowance and subsequently giving the proceeds from the transaction to the recipient.[104]

The Joint Committee on Taxation has added that considering emission allowances to be tradable, and thus "cash-like," makes them similar to commodities, noting that sulfur dioxide and nitrogen oxide emission allowances created by the Clean Air Act and various types of carbon credits and their derivatives are already traded on commodities markets.[105] At the same time, the Committee found that allowances also "bear some resemblance to licenses that the government grants in other contexts, e.g., television broadcast licenses granted by the Federal Communication Commissions, liquor licenses granted by State and local governments, and certain agricultural production quotas."[106] As with these licenses, "emission allowances are transferable, intangible assets, the useful life of which can be limited by statute."[107] The Committee continued:

> The application of different analogies can lead to very different answers to the most basic tax questions presented by cap and trade. For example, whereas allocations of certain licenses by the government have been deemed to be nonrecognition events (i.e., no tax is imposed at the time the license is granted), few would argue that a government distribution of a commodity, such as gold, oil, or pork bellies, should not be taxable to the recipient.[108]

The Committee identified three alternatives for taxing allocated emission allowances, based on whether and when there would be an accession to wealth: (1) including them in income upon receipt, (2) including them in income when first available for use, and (3) excluding them from income.[109]

It is clear that in providing emission allowances to domestic entities, the federal government would provide the recipient with a vehicle for the receipt of a monetary benefit, and thus it may be viewed in a broad sense as providing a subsidy to the recipient entity. As explained below, however, for the provision of emission allowances to constitute a subsidy in WTO terms, the government activity or practice must first qualify as a "financial contribution" or "an income or price support" as those terms are understood under WTO agreements. To date, neither GATT nor WTO jurisprudence has addressed this type of instrument in light of WTO subsidy obligations.

WTO Agreement on Subsidies and Countervailing Measures (SCM)

The provision of subsidies by WTO Members is governed by the WTO Agreement on Subsidies and Countervailing Measures (SCM Agreement), which elaborates upon and expands subsidy obligations contained in Article XVI of the GATT 1994 and contains detailed obligations involving the imposition of countervailing duties permitted under GATT Article VI. WTO Members may impose countervailing duties on imported products that are found to be subsidized by an exporting WTO Member and cause or threaten material injury to (or materially retard the establishment of) a domestic industry.[110] A subsidy meeting the WTO definition may be challenged in a WTO dispute settlement proceeding or may be remedied by the imposition of countervailing duties on the subsidized product in an amount that does not exceed the subsidy conferred. While the GATT 1994 contains general public policy-related exceptions that may be invoked to justify GATT-inconsistent measures, the SCM Agreement does not contain a separate set of exceptions that would permit WTO Members to deviate from agreement obligations.[111]

For purposes of the SCM Agreement, the term subsidy is defined as a "financial contribution by a government or any public body within the territory of a Member," or an income or price support in the sense of Article XVI of the GATT 1994 that confers a benefit.[112] A financial contribution will be found where:

> (i) a government practice involves a direct transfer of funds (e.g., grants, loans, and equity infusion), potential direct transfer of funds or liabilities (e.g., loan guarantees);
>
> (ii) government revenue that is otherwise due is foregone or not collected (e.g., fiscal incentives such as tax credits);
>
> (iii) a government provides goods or services other than general infrastructure, or purchases goods;
>
> (iv) a government makes payments to a funding mechanism, or entrusts or directs a private body to carry out one or more of the functions illustrated in (i) to (iii) above which would normally be vested in the government and the practice, in no real sense, differs from practices normally followed by governments.[113]

While an income or price support may constitute the requisite governmental involvement for purposes of the SCM Agreement, this provision has not been cited to any great extent in GATT or WTO jurisprudence.[114] With respect to the second prong of the WTO definition, that is, the conferral of a benefit, a financial contribution will be found to do so if it places the recipient in a more advantageous situation than would have been the case absent the contribution.[115]

To be challenged in a WTO dispute settlement proceeding or to be subject to countervailing duties, the subsidy must be specific to an industry or enterprise or a group of industries or enterprises.[116] Prohibited subsidies, as described below, are considered to be specific *per se*.[117] Subsidies may be specific in law, that is, they may be explicitly limited to certain enterprises and not be administered under objective criteria or conditions, or they may be specific in fact.[118] Regarding the rules under which the program operates, the SCM Agreement provides that specificity will not exist where legislation, or the granting authority operating under it, "establishes objective criteria or conditions governing the eligibility for, and the amount of, a subsidy ... provided that the eligibility is automatic and that such criteria and conditions are strictly adhered to."[119] Objective criteria or conditions mean those "which are neutral, which do not favour certain enterprises over others, and which are economic in nature and horizontal in application, such as number of employees or size of enterprise."[120]

The SCM Agreement divides subsidies into two categories: prohibited and actionable. Two types of subsidies are prohibited: (1) subsidies "contingent, in law or in fact ... upon export performance" and (2) subsidies "contingent ... upon the use of domestic over imported products" (also referred to as "import substitution" subsidies).[121] The mere fact that a subsidy is granted to a firm that exports is not enough to render it an export subsidy for purposes of the Agreement.[122]

Subsidies fitting the WTO definition that are not prohibited are considered "actionable," that is, they may be challenged in a WTO dispute settlement proceeding if they cause "adverse effects" to the interests of another WTO Member.[123] Under Article 5 of the Agreement, adverse effects may take any of three forms: (1) injury to the domestic industry of another Member, as this concept is used in countervailing duty proceedings (a standard that focuses on the effect of the subsidized goods in the domestic market of the complaining Member); (2) nullification or impairment of another Member's WTO benefits, generally tariff concessions on a given product; and (3) serious prejudice to the Member's interests.

As set out in Article 6.3 of the SCM Agreement, serious prejudice occurs when the effect of the subsidy is (1) to displace imports of a like product of the complaining Member into the market of the subsidizing Member; (2) to displace or impede the exports of a like product of the complaining Member from a third country market; (3) significant price undercutting by the subsidized product as compared with the price of a like product of the complaining Member in the same market, or significant price suppression, price depression, or lost sales in the same market; and (4) an increase in the world market share of the subsidizing Member in a particular subsidized primary product or commodity as compared to the average share that the subsidizing Member had during the previous period three-year period and the increase follows a consistent trend over a period when subsidies have been granted. [124] In any such case, defining the nature of the "like product" and the affected market would be important components in determining if serious prejudice exists.[125]

Under special dispute settlement rules for the SCM Agreement, if the WTO Dispute Settlement Body adopts a panel or Appellate Body report finding that a subsidy has resulted in adverse effects to another Member, the subsidizing Member "shall take appropriate steps to remove the adverse effects or shall withdraw the subsidy."[126] If the Member has not done so within six months after adoption, and absent an agreement on compensation, the Dispute Settlement Body is to authorize the complaining Member to take countermeasures, "commensurate with the degree and nature of the adverse effects determined to exist," unless the Dispute Settlement Body decides by consensus to reject the complaining Member's request to impose such measures.[127] This time period may be extended by mutual agreement of the disputing parties.[128]

Free Allowances under the SCM Agreement

For the provision of emission allowances to fit within the SCM Agreement's definition of a governmental financial contribution, the action would need to constitute (1) an actual or potential direct transfer of funds, (2) the foregoing of revenue otherwise due, or (3) the provision of a good or service other than general infrastructure. Because WTO panels have not had to deal with an instrument of this type, it is unclear how it would or should be characterized under this definition. It is also unclear how domestic tax or budgetary treatment of an allowance might affect this characterization.[129] The precise nature of an allowance is elusive for these purposes, and thus a variety of scenarios can be contemplated. This chapter addresses some of the more salient subsidy issues that may arise in this context.

While emissions have been characterized as "cash-like" in nature and would clearly constitute a valuable instrument from the point of view of the recipient, to the extent that an allowance is intended to be sold or traded, the allowance would in and of itself constitute a vehicle for a financial contribution by private parties, that is, the ultimate transfer of funds would be effected by the purchasers of the emission allowance rather than by the government. In such case, the transfer may not be the type of "direct transfer" of funds by the government that is generally contemplated by the first type of financial contribution listed above.[130]

As an emission allowance has also been characterized as a license and a commodity, one might alternatively argue that the provision of an allowance constitutes the provision of a good. While the WTO Appellate Body has confirmed that the granting of a government license may constitute the provision of goods to a recipient, its finding would appear to have limited utility in the current context. Because the license at issue permitted recipients to harvest standing timber on government lands, the government grant of a license was thus found to constitute the provision of timber, a potentially tradable product.[131] In contrast, the provision of a free allowance would represent permission or authority to emit a defined amount of carbon dioxide or a carbon dioxide equivalent, a substance that would not be a salable good in the same sense as timber was in the above-cited example. Thus, if a similar analysis is applied to emission allowances, this category of government financial contribution is not likely to apply.[132] Further, unlike commodities that are in and of themselves tradable goods, the item that is being traded here would essentially be a right to take a particular action rather than a tangible product.[133]

A case for a subsidy may be made, however, once emission allowances are subject to government auction, an event contemplated by H.R. 2454 to begin in 2012. Under the SCM Agreement, the concept of revenue that is "otherwise due" requires an ascertainable standard against which a tax or other exemption is measured. As described by the WTO, this portion of the subsidy definition implies "an understanding that (i) 'a financial contribution' does not arise simply because a government does not raise revenue which it could have raised; and (ii) the term 'otherwise due' implies a comparison with a 'defined normative benchmark.'"[134] In such case, the provision of an allowance without charge to a U.S. firm may arguably constitute the foregoing by the government of revenue that would otherwise be due, the specifics of the government auction serving as the applicable norm. It is also possible that the future tax treatment of distributed allowances may itself result in such foregone revenue. Although WTO jurisprudence on this portion of the subsidy definition most often focuses on tax measures,[135] the provision itself is generally written and is not limited to the tax area.[136]

In the event that the provision of free allowances or the tax rules that applied to them were found to qualify as foregone revenue, the existence of a benefit may, in some situations, not be difficult to discern. In past cases, WTO panels, in determinations not subsequently appealed, have found that the tax exemptions were virtually coterminous with the existence of a benefit.[137] For example, in a report involving tax exemptions under the U.S. Foreign Sales Corporation (FSC) statute, the panel stated that, in its view, "the financial contribution clearly confers a benefit, in as much as both FSCs and their parents need not pay certain taxes that would otherwise be due," noting further that the United States had not raised any contrary arguments regarding the benefit issue.[138] Similarly, in a subsequent case that involved a statute that repealed the FSC statute and established a replacement tax regime permitting the exclusion from taxation of certain income, the panel stated that under the new statute:

> a taxpayer involved in a qualifying transaction may exclude qualifying foreign trade income from its gross income and therefore need not pay a certain amount of tax that it would otherwise have to pay to the United States government. It is therefore "better off" than it would have been absent the contribution, that is, if had been in another situation, where the conditions for obtaining the tax treatment under the Act were not fulfilled and it was therefore subject to otherwise applicable US taxation rules. We are of the view that the tax treatment in the Act confers a benefit.[139]

Similarly, in a case involving whether an exemption from an import duty constituted a subsidy, the WTO panel, having determined that the exemption qualified as revenue foregone, easily found that it conferred a benefit since "the fact that manufacturer beneficiaries need not pay customs duties that would otherwise be due—and that would be paid by non-qualifying manufacturers—constitutes ... an advantage" of the type that the Appellate Body had found in an earlier case.[140]

In an assessment of whether a benefit is conferred, panels must focus on the recipient of the financial contribution rather than on the granting authority and the cost to the granting government.[141] Because of the various situations of recipients of free allowances and because there may be particular legislative requirements or conditions accompanying their receipt, the existence or nature of the benefit, that is, whether a particular recipient is "better off," may vary with respect to the recipient involved. Since WTO panels have not yet dealt with such an instrument, this remains a matter for further development by WTO panels and the WTO Appellate Body.

Were the provision of free allowances to constitute a subsidy for purposes of the SCM Agreement, the subsidy would not be prohibited under the Agreement so long as provision of the allowances was not contingent in law or in fact on export performance or on the use of domestic over imported products. Nevertheless, the subsidy would potentially be actionable, and thus subject to challenge if a complaining WTO Member could show that it had suffered one of the adverse effects set out in Article 5 of the Agreement. The complaining Member would also need to show that the subsidy was specific in law or fact to an industry or group of industries before proceeding with its showing of trade injury. While free allowances may be available to entities in a broad range of economic sectors and might be viewed as not limited by statute to "certain enterprises," determining whether "objective criteria and conditions" governing the eligibility for, and the amount of, the subsidy exist would appear to be an important area of inquiry in determining if specificity is present.

International Reserve Allowance (IRA) Program

H.R. 2454, as passed, would also require that, if the President established an IRA program for an eligible domestic industrial sector and once he determines that the 85% import threshold for the sector is not exceeded, importers of products corresponding to those produced by the sector submit emission credits upon the entry of these products into the United States. The requirement would apply to a particular importation unless EPA has adjusted the border allowance requirement to zero for the product or the product is imported from a statutorily exempted country. In addition, "manufactured items for consumption," that is, downstream products containing inputs that are subject to sectoral IRA requirements, may also be covered by border IRA requirements. Exemptions would apply to products imported from countries that have entered into acceptable multilateral or bilateral GHG-reduction agreements or whose annual energy or GHG-intensity for the sector is equal to or less than that of the United States. Also exempted would be products imported from the least developed of developing countries and from countries that are *de minimis* emitters and are the source of less than 5% of U.S. imports of covered goods for a sector. In issuing regulations establishing the program, EPA , with the concurrence of U.S. Customs and Border Protection, would incorporate certain statutory requirements, but would also be given discretion to establish a general methodology for calculating the quantity of IRAs that a U.S. importer of any covered good must submit and would be authorized to adjust IRA requirements based on emission allowances distributed to eligible industrial sectors and to reduce the amount to zero. The IRA requirements would not apply to goods that enter the United States before January 1, 2020, the earliest date for application specified in the statute.

General Agreement on Tariffs and Trade (1994)

Article I, the general most-favored-nation (MFN) obligation of the GATT, requires that certain trade-related benefits that a WTO Member grants to the products of any country must be granted "immediately and unconditionally" to like products of all WTO Members. The obligation applies to any advantage involving customs duties and charges of any kind imposed on importation or exportation, the method of levying such duties, all rules and formalities connected with importation and exportation, and all matters related to internal taxation and regulation, that is, matters covered by the GATT national treatment article.

Article II, which generally prohibits tariff surcharges and border fees on imports, is aimed at ensuring that tariff concessions negotiated by WTO Members are maintained at negotiated rates. Each Member's tariff commitments are set out in a Schedule which lists the highest rate that a Member may impose on a given product, also known as the "bound" rate. Article II:1(b) provides that products imported from other WTO Members are to be exempt from ordinary customs duties in excess of those set out in the importing Member's Schedule and from "all other duties and charges of any kind imposed on or in connection with importation in excess of those imposed" on April 15, 1994, the date the WTO agreements were concluded. At the same time, Article II:2(a) provides that Article II does not prevent a WTO Member from imposing specific types of charges on bound items, including "a charge equivalent to an internal tax imposed consistently with the provisions" of the GATT national treatment article "in respect of the like domestic product or in respect of an article from which the imported product has been manufactured or produced in whole or in part."

Article III, the GATT national treatment article, prohibits WTO Members from discriminating between like domestic and imported products when imposing internal taxes and regulations. Where an internal tax or regulation is enforced at the border with regard to an imported product, the tax or regulation will be considered an internal measure and, as such, subject to Article III.

Article III:1 of the GATT 1994, which generally informs Article III obligations, states that "internal taxes and other internal charges and laws, regulations and requirements affecting the internal sale, offering for sale, purchase, transportation, distribution or use of products ... should not be applied to imported or domestic products so as to afford protection to domestic production."

Article III:2, which addresses taxes on products (e.g., excise and sales taxes, also referred to as "indirect taxes") states that the products of a WTO Member imported into the territory of another WTO Member "shall not be subject, directly or indirectly to internal taxes or other internal charges of any kind in excess of those applied, directly or indirectly, to like domestic products." To determine whether the tax on an import exceeds the tax on the like domestic product, a strict test is applied, under which "even the smallest amount of 'excess' is too much"; neither a "trade effects" test nor a *de minimis* standard qualifies the prohibition. [142] Further, Article III:2 requires that actual, rather than nominal, tax burdens be compared. An identical tax rate can be found in some cases to result in a heavier tax burden on an import because of the method of taxation, and thus a WTO review would likely take into account not only the tax rate but also "the taxation methods (e.g., different kinds of internal taxes, direct taxation of the finished products, or indirect taxation by taxing the raw materials used in the product during the various stages of its production) and of the rules for tax collection (e.g., the basis of assessment)."[143] In general, panels have viewed the policy purpose behind a tax as irrelevant so long as the Member imposing the tax does not violate the GATT or other WTO obligations.[144]

Article III:4 of the GATT 1994, requiring national treatment in internal regulation, states that products of any WTO Member imported into the territory of any other Member "shall be accorded treatment no less favourable than that accorded to like products of national origin in respect of all laws, regulations and requirements affecting their internal sale, offering for sale, purchase, transportation, distribution or use."

In determining whether imported and domestic goods are "like" products for purposes of both GATT Article III:2 and Article III:4, panels have ordinarily used four criteria: (1) the properties, nature, and quality of the products; (2) end-uses; (3) consumers' tastes and habits; and (4) tariff classification. [145] To date, GATT/WTO case law has not permitted Members, under GATT Article III, to distinguish between products that would otherwise be considered "like" on the basis of a processing and production method (PPM) that does not relate to a product characteristic. Thus, applying such a distinction in a regulatory scheme and subjecting products to different requirements based on this distinction may provide a basis for finding that an imported good is treated less favorably than the like domestic item or that any resulting import prohibition constitutes a quantitative restriction prohibited under Article XI, below.[146]

Article XI: 1 of the GATT generally prohibits quantitative restrictions on imports and exports, providing that "[no] prohibitions or restrictions other than duties, taxes or other charges, whether made effective through quotas, import or export licenses or other measures, shall be instituted or maintained by any contracting party [i.e. WTO Member] on the importation of any product of the territory of any other contracting party or on the exportation or sale for export of any product destined for the territory of another contracting party." Deviations from this rule are allowed only in certain well-defined circumstances, generally unrelated to climate change issues. A quantitative restriction may be distinguished from an internal regulation enforced at the border by examining whether the measure affects the opportunity for importation or entering the market, in which case Article XI would apply, or whether it affects competitive opportunities in the domestic market, in which case Article III would govern.[147]

Article XX, containing the GATT general exceptions and operating as a defense in GATT disputes, allows a WTO Member to justify, on a variety of public policy grounds, a measure that has been found to violate a GATT obligation. Conditioning market access on compliance with a policy unilaterally prescribed by a WTO Member, while potentially irksome to trading partners, has been viewed by the WTO Appellate Body as a common feature of measures falling within the scope of Article XX exceptions, and thus such a policy may be pursued under the GATT providing implementing measures satisfy Article XX requirements.[148]

Of relevance in the climate change context are ***Article XX(b)***, covering measures "necessary to protect human, animal or plant life or health" and ***Article XX(g)***, covering measures "relating to the conservation of exhaustible natural resources if such measures are made effective in conjunction with restrictions on domestic production or consumption." Any law or regulation that is provisionally justified under Article XX(b) or Article XX(g) must also comply with the proviso to Article XX, which requires that the measure not be applied "in a manner which would constitute a means of arbitrary or unjustifiable discrimination between countries where the same conditions prevail, or a disguised restriction on international trade."

Border IRA Requirements under the GATT 1994

As explained earlier in this chapter, the manufacture of certain products results in direct carbon emissions. Production of these and other products may also result in what are deemed to be indirect carbon emissions due to the high amount of energy (generally electrical energy) needed for their production and the heavy use of carbon-based fuels to provide such energy. Thus, a focus of concern in a domestic GHG-reduction program is its implications for goods that are carbon- intensive, energy-intensive, or both.

Where emission caps are placed on energy producers and on manufacturers of carbon-intensive products, the cost of the caps for these producers and manufacturers, as well as for manufacturers that are heavy users of carbon-intensive energy, may adversely affect the competitiveness of these manufacturers' products *vis à vis* the same or similar products produced in countries without carbon controls. Depending on the situation of the firm and the products produced, the loss of competitiveness may occur on the import side, the export side, or both. A significant and irremediable loss of competitiveness may cause such firms to move their production to countries without GHG-reduction controls, potentially resulting in increased GHG emissions in these countries, or "carbon leakage." Increased emissions may also result from increased production by existing foreign facilities whose presumably more price-competitive goods would be in greater demand worldwide. In such case, a loss of competitiveness, an economic concern, would be the reason for an adverse environmental effect. A border IRA requirement that takes into account the level of greenhouse gases emitted in the production of the imported product in light of domestic regulatory requirements may seek to "level the playing field" for a particular product or sector and so prevent the carbon leakage that may negate the beneficial environmental outcome of the domestic GHG-reduction program.[149]

Requirements of H.R. 2454 and GATT Articles

A border requirement of this type proposed under H.R. 2454—that is, requirement that an importer of a product from a country that has not taken sufficient action to reduce GHG emissions, as judged by the importing country—potentially implicates the GATT articles described above, namely, Article I, requiring most-favored-nation treatment; Article II, prohibiting added duties and other fees and charges on goods subject to negotiated tariff rates; Article III, requiring national treatment of imported products; and Article XI, prohibiting quantitative restrictions on imports.

Since an importer of a foreign-produced item would incur a cost in amassing the IRAs that would be needed for importation under H.R. 2454, the fees and charges paid by the importer may be found to constitute a "charge of any kind imposed on or in connection with the importation" of a item subject to tariff rates bound under GATT Article II and, as such, be prohibited under Article II:1(b). Because the vast majority of products in the U.S. GATT Schedule are subject to bound tariff rates, the requirements of Article II would apply broadly to products imported into this country. In one instance, a GATT panel found that the interest charges and other costs connected with the posting of security to guarantee that imports of a particular item would be made at a minimum price were "'other duties or charges of any kind imposed on or in connection with the importation' in excess of the bound rate within the meaning of Article II:1(b)," and thus inconsistent with this obligation.[150] More recently, a WTO panel, citing this GATT case, found that the interest charges, costs, and fees incurred by importers in connection with posting an additional customs bond required by the United States were import charges prohibited by the above-quoted Article II:1(b) language. [151] The United States did not appeal this finding.[152] As in these cases, the importer subject to the IRA requirements would incur the cost of obtaining the allowances to satisfy a requirement associated with an importation rather than pay a specified fee imposed directly on the carbon-intensive or energy-intensive product. Since Article II:1(b) speaks broadly of "duties or *charges of any kind*," the expenditure of funds to obtain the required allowances may well result in the type of charge that falls within the scope of the Article II:1(b) obligation.

If the importer's failure to comply with such requirements serves to prohibit a product from being imported into the United States, this aspect of the IRA program may also be viewed as inconsistent with the general prohibition on quantitative restrictions in GATT Article XI: 1. Further, to the extent that a fee or charge would not apply to goods originating in a country with a GHG-reduction program or in a country exempted for other reasons (e.g., *de minimis* emissions), there may arguably exist discriminatory treatment of like products from non-exempted countries for purposes of the most-favored-nation obligation of GATT Article I.

Alternatively, the United States may be able to avoid the Article XI prohibition on quantitative restrictions if the IRA requirement on imports could be shown to be part of an internal regulatory regime governing the "internal sale, offering for sale, purchase, transportation, distribution, or use" of carbon-intensive and energy-intensive products for purpose of GATT Article III:4. If so, the United States may be able to prohibit imports that do not meet U.S. regulatory requirements from entering the United States without violating

Article XI. It appears difficult, however, to fit a border program such as provided for in H.R. 2454 within the parameters of this GATT article.

Generally speaking, the regulation of products based on their carbon emissions raises issues as to whether otherwise like items (e.g, a particular type of steel product) may be distinguished on this basis under current WTO law. If steel products are found to be like products based on the criteria ordinarily used by panels (i.e., product characteristics, end uses, consumer preference, and tariff classification), distinguishing them based on a substance emitted in their production may thus be a problematic basis for differing regulatory requirements, and thus a case for less favorable treatment of the like imported good could be made. More fundamentally, however, because H.R. 2454 would place its relevant domestic requirement (i.e., emission caps, on producers and not on the sale, purchase, or use of domestically produced carbon-intensive and energy-intensive goods), the bill would not appear to create the sort of domestic regulatory regime affecting domestic and imported *products* that is essentially contemplated under Article III. Thus, even though the production of domestic carbon-intensive and energy-intensive products may be economically affected by the requirements placed on domestic producers, the sale, purchase, or use of these goods would not be subject to regulatory requirements. The absence of a counterpart internal regulatory program applicable to goods produced in the United States would thus render the import prohibitions under the IRA program liable to the Article XI claims discussed above.

Justification of H.R. 2454 IRA Program under Article XX General Exceptions

As noted above, if a trade-related GHG reduction measure is challenged in a WTO dispute settlement proceeding and found to violate a GATT obligation, the defending Member may seek to justify it under a GATT general exception, the most likely candidates in the climate change context being Articles XX(b), covering measures "necessary to protect human, animal, or plant life or health" and Article XX(g), covering measures "relating to the conservation of exhaustible natural resources if such measures are made effective in conjunction with restrictions on domestic production or consumption."[153] A measure falling within the scope of an exception is also subject to the overall Article XX proviso or "chapeau" requiring that any such measure not be "applied in a manner which would constitute a means of arbitrary or unjustifiable discrimination between countries where the same conditions prevail or a disguised restriction on international trade."

Finding that a measure is justified under an exception involves a complex analysis of the alleged policy goal, the relationship of the measure to the goal, and details of the measure's implementation in light of the discriminatory and protectionist application to be avoided under the proviso. If challenged, it would not be the GATT-inconsistency that would need to fall within the scope of the exception, but rather the border measure program as a whole.[154] Because Article XX(b) requires a showing that a challenged measure is "necessary" to achieve the aim of health protection, whereas Article XX(g) requires only that a relationship be shown between the challenged measure and the stated conservation goal, it would generally be more difficult for a measure to qualify under the former than under the latter. In either case, however, since maintaining domestic competitiveness *per se* is not a policy goal protected by Article XX, the extent to which the IRA requirements of H.R. 2454 address articulated goals of protecting health or conserving natural resources would be critical in determining whether the program falls within the scope of the relevant exception.

Article XX(b)

Successful invocation of Article XX(b) first requires a showing that the policy objective of the challenged measure is to protect human, animal, or plant life or health, and that the measure is necessary to achieve this end. As noted below, the WTO Appellate Body has implied that health protection may be a goal of a domestic climate change program. At the same time, since the purported health protection would be achieved through the prevention of carbon leakage that is manifested as increased carbon emissions in foreign countries, and because Article XX(b) does not expressly state whether the objects of protection need to be located in the territory of the Member imposing the restriction, a jurisdictional issue may arise in a climate change case, requiring a panel to clarify Article XX(b) 's territorial reach and how it may be satisfied. The global impact of the climate change problem may be sufficient, however, to find any local nexus that may be needed.[155]

To find necessity, the Appellate Body has articulated a test that includes weighing and balancing of "the relevant factors" with an examination of less trade-restrictive alternatives. Under this process, the panel (1) considers "the importance of the interests or values at stake, the extent of the contribution to the achievement to the measure's objective, and its trade restrictiveness," and (2) if the panel preliminarily finds necessity, "confirms" this conclusion "by comparing the measure with possible alternatives that may be less trade restrictive while providing an equivalent contribution to the achievement of the objective." [156] The comparison "should be carried out in the light of the importance of the interests or values at stake."[157]

The WTO Appellate Body has stated that the more vital or important the value being pursued, the easier it would be to find that the chosen measure is necessary to achieve the chosen level of health protection.[158] The measure does not need to be shown to be indispensable, but, under recent WTO jurisprudence, it must be "apt to produce a material contribution to the achievement of its objective"; in other words, a measure providing a "marginal or insignificant" contribution would not be considered necessary to achieve the stated goal.[159] Immediate impact of the measure need not be shown, the WTO Appellate Body having recognized that solving complex environmental problems may require a range of interacting measures and that "the results from certain actions—for instance, measures adopted in order to attenuate global warming and climate change ...—can only be evaluated with the benefit of time."[160] A higher level of contribution to achieving the stated goal may be needed where the measure is particularly restrictive, as is the case with an import ban.[161] It would be up to the complaining Member in a dispute to identify possible alternative measures; if it does so, the defending Member would be able to rebut such suggestions on the grounds that the alternative inadequately contributes to the goal or is not reasonably available for this purpose.

Article XX(g)

For Article XX(g) to apply to a GATT-inconsistent climate change measure, the WTO Member would need to show (1) that conservation of an exhaustible natural resource is at issue; (2) that the measure relates to conservation of this resource; and (3) that the measure is made effective in conjunction with restrictions on domestic consumption or production. In *United States–Import Prohibition of Certain Shrimp and Shrimp Products* (*U.S. Shrimp*), the WTO Appellate Body took a broad view of exhaustibility in light of evolving multilateral agreement and action in the conservation area since the adoption of the GATT in 1947 and found that the Article XX(g) applies not only to exhaustible mineral or other non-living natural resources, but also to all exhaustible resources, whether living or non-living. [162] Since the renewability of a resource would not preclude it from falling within the scope of the exception,[163] the exhaustible resource at issue in the climate change context might be the atmosphere at a suitable temperature or a species adversely affected by rising global temperatures. As with Article XX(b), however, the jurisdictional reach of the provision may need to be addressed, with similar considerations coming into play.[164]

For a measure to "relate to" the conservation of exhaustible natural resources, it must be "primarily aimed at" this goal.[165] As articulated in *U.S. Shrimp*, this test requires a "substantial relationship" between "the general structure and the design of the measure ... and the policy goal it purports to serve," a situation also characterized as "a close and genuine relationship of ends and means."[166] The relationship of a climate change measure to the claimed conservation goal is key, thus requiring a focus on how the measure would prevent carbon leakage and therefore conserve the earlier-identified natural resources. Evolving studies on the prevention of leakage may, however, present problems in this regard. Even though the WTO Appellate Body has indicated that the immediate effect of a GATT-inconsistent climate change measure may not need to be established in order to successfully invoke a GATT exception, the existence of credible studies questioning whether leakage would be prevented by such measures may increase the difficulty of showing that the program is primarily aimed at preserving the natural resource or resources shown to be at risk.

To show that the measure is made effective in conjunction with restrictions on domestic consumption or production, a panel would examine whether the restriction on the imported product is imposed with respect to the same domestic items and whether the restriction, while not needing to provide true equality of treatment between the two,[167] is "even-handed" in its approach to imports *vis à vis* the restriction placed on domestic goods.[168] The existence of "evenhandedness" may become an issue to the extent that H.R. 2454 permits border measures to apply to "manufactured items for consumption" or the imported counterparts of items produced by industrial sectors that did not initially qualify for the receipt of emission allowances under the statute.

Article XX "chapeau"

The Article XX "chapeau," which is aimed at preventing abuse of the Article XX exceptions, focuses on how the GATT-inconsistent measure is applied. In the view of the WTO Appellate Body, interpreting and applying the proviso is a "delicate" task of finding "a line of equilibrium" between the right of a Member to invoke an Article XX exception and the rights of other Members under GATT substantive obligations; the line moves "as the kind and the shape of the measures at stake vary and as the facts making up specific cases differ."[169] In a 2007 case, *Brazil—Measures Affecting Imports of Retreaded Tyres* (*Brazil Tyres*), the WTO Appellate Body examined earlier cases in which it has applied the proviso and determined that the common mode of analysis in cases involving arbitrary or unjustifiable discrimination involved a determination as to whether the discrimination "had a legitimate cause or rationale in light of the objectives listed in the paragraphs of Article XX."[170] The second element of the proviso has been the subject of less jurisprudence, but it appears to be agreed upon that the prohibition on creating a "disguised trade restriction" is aimed at avoiding a protectionist effect.

Unjustifiable discrimination was found to arise in two aspects of the U.S. pollution reduction program at issue in *United States—Standards for Reformulated and Conventional Gasoline* (*U.S. Gasoline*), a case challenging the application of a statutory baseline to foreign refiners in assessing whether imported gasoline met Clean Air Act standards, while more favorable individual baselines were applied to their U.S. counterparts. First, the United States had failed to engage affected exporting countries in exploring cooperative arrangements to mitigate administrative difficulties that the United States claimed would exist in acquiring foreign data for verification and assessment purposes if individual foreign baselines were used.[171] Second, the United States had taken into account burdensome costs that would be placed on domestic refineries if they too were subject to the statutory baseline, but had not considered costs that would be incurred by foreign firms under the program.[172]

In *U.S. Shrimp*, the Appellate Body found unjustifiable discrimination in the "intended and actual coercive effect" on foreign government policy decisions of the U.S. program prohibiting the importation of shrimp from countries not certified by the United States as maintaining a regulatory program and an incidental taking rate of sea turtles comparable to that of the United States or having a fishing environment that did not pose a threat to sea turtles.[173] According to the Appellate Body, the United States had implemented the statute to require that exporting countries adopt a regulatory program with policies and enforcement practices that were "not merely *comparable*, but rather *essentially the same,*" as the program applied to U.S. shrimp trawlers, and in some cases, to prohibit the importation of shrimp caught with methods identical to those employed in the United States because the shrimp were harvested in waters of countries not certified by the United States as using acceptable techniques.[174] The Appellate Body stated that discrimination results "not only when countries are differently treated, but also when the application of the measure at issue does not allow for any inquiry into the appropriateness of the regulatory program for the conditions prevailing in those exporting countries."[175]

As in *U.S. Gasoline*, the Appellate Body in *U.S. Shrimp* also found unjustifiable discrimination in the failure of the United States to engage affected WTO Members diplomatically—here, "serious across-the-board negotiations with the objective of concluding bilateral or multilateral agreements for the protection and conservation of sea turtles"—before enforcing the import prohibition on shrimp against these countries.[176] Actual conclusion of an international agreement is not required, however, so long as "serious, good faith efforts" are made to negotiate an accord.[177]

The Appellate Body has also indicated that it will look at the length of the phase-in period for foreign compliance with a regulatory program, both as to compliance burdens resulting from a relatively short period and any differences in treatment between affected countries.[178]

Arbitrary discrimination may likewise result from placing a "single, rigid, unbending requirement" on foreign countries under a regulatory program, as well as from not according agency officials sufficient flexibility in making determinations under it.[179] The Appellate Body looked for flexibility in implementation, stating that where market access is conditioned on an exporting Member adopting a regulatory program that is "comparable in effectiveness" to that of the importing Member, the exporting Member will have "sufficient latitude ... with respect to the programme it may adopt to achieve the level of effectiveness required" and may thus adopt a program that is "suitable to the specific condition prevailing in its territory."[180]

Avoiding arbitrary discrimination has also been found to implicate due process concerns, and thus, where a statute requires that exporting countries fulfill certain conditions before their exports are permitted entry into the United States, the accompanying regulatory process should be "transparent" and "predictable " and should not be *ex parte,* that is, conducted only with input from the importing Member's agencies and officials. [181] Instead, the importing Member should give affected exporting countries an opportunity to explain their situation to agency decision-makers, and provide them with a reasoned explanation for denial of requests and petitions, as well as procedures for review of any such denials.[182]

Determining whether a disguised restriction on international trade exists involves focusing not on the restriction *per se*, but rather on what it may mask. Since this portion of the Article XX proviso is also aimed at avoiding abuse and illegitimate use of the Article XX exceptions, the Appellate Body has found that a restriction that formally meets the requirements of an exception "will constitute an abuse if such compliance is in fact only a disguise to conceal the pursuit of trade- restrictive objectives."[183] Because the aim of a measure "may not be easily ascertained ... the protective application of a measure can most often be discerned from its design, architecture, and revealing structure."[184] While the Appellate Body has indicated that the same considerations that are used to determine if discrimination exists may be used to find a disguised trade restriction,[185] a subsequent WTO panel chose not to examine whether discrimination existed for this portion of the proviso where discrimination had not already been found. Instead, it focused on possible protectionist objectives of the measure and the extent to which it had benefited a domestic industry to the detriment of foreign producers.[186]

Given these cases, were the IRA program found to violate U.S. GATT obligations but the program was found to fall within the scope of a GATT general exception, the United States would seemingly need to address issues such as the following to show that the program was applied consistently with the Article XX proviso: (1) engagement of trading partners for the purpose of negotiating a bilateral or multilateral solution both to the GHG-reduction goal and addressing the use of border measures to remedy the competitiveness concerns that lead to leakage and administrative problems that may arise in implementing the border measure program, particularly in regard to the use of foreign data; (2) use of a flexible regulatory standard under which the Member permits comparability in the effectiveness of foreign programs in achieving the importing Member's policy goal; (3) creation of a regulatory process that gives exporting Members a meaningful opportunity to be heard and to resolve problems and deficiencies in seeking and obtaining access to the importing Member's market; (4) providing an adequate phase-in period before import requirements enter into effect; and (5) permitting the importation of goods that in fact comply with the requirements of the regulatory program, thus avoiding overly broad regulatory categories that may result in penalizing such compliance.

As discussed earlier, the negotiating objectives under Part IV, Subpart 2 include seeking both multilateral GHG-reduction commitments and provisions permitting parties to remedy competitive imbalances that lead to leakage. Additionally, however, seeking data cooperation agreements may also be important in that the regulatory program may rely heavily on foreign emissions and production statistics. The avoidance of arbitrary or unjustifiable discrimination may depend on the quality of such data where they are used to assess emissions-reduction activity in order to determine whether and to what extent border adjustment apply, as well as to distinguish between countries "where the same conditions prevail." Arguably, even though modeling may potentially be used under the legislation to produce otherwise unobtainable hard data, a general lack of usable emissions data may raise questions as to whether the border requirement may be applied in a nondiscriminatory fashion in all cases for purposes of Article XX.

Further, even assuming that reliable foreign data existed, problems may nevertheless arise in obtaining it. In the absence of a WTO agreement, such as the WTO Antidumping Agreement and the Agreement on Subsidies and Countervailing Measures (under which WTO Members have agreed that they or their exporters will provide information to authorities of the importing Member for investigatory or regulatory purposes), exporting countries and their firms may be reluctant to provide emissions or production data to EPA for purposes of making comparability or other determinations under the IRA program. Unlike the situation in *U.S. Gasoline*, where the quality of the imported gasoline directly affected the quality of the atmosphere within the United States, the carbon-intensive and energy-intensive goods that would be subject to IRA requirements would not themselves be environmentally harmful within U.S. territory. In such case, foreign entities may arguably feel less compelled to produce data than they would in a situation where their products clearly contributed to a problem within the sovereign jurisdiction of the United States.

The issue of penalizing compliant imports may arise where the importing country measures the comparability of a foreign program by determining an average national emissions rate, bases the required amount of border emission credits for an imported product on this average, and subjects all such goods originating in the country to the same requirement, regardless of actual GHG gases emitted in their production. Other issues are likely to arise where import requirements are based on GHG emissions, for example, accurately determining the level of emissions attributable to production of goods occurring in multiple countries under a variety of production processes. In addition, the extent to which a border program covers imported goods that are downstream from the type of products that are produced by domestically capped manufacturers or by initially eligible industrial sectors could be a factor in examining whether the program is applied in a manner that constitutes a disguised restriction on international trade.

APPENDIX. EC LIST OF ELIGIBLE INDUSTRIES

1. At the NACE-4 level

1.1. Based on the quantitative criteria set out in paragraphs 15 and 16 of Article 10a of Directive 2003/87/EC

NACE Code	Description
1010	Mining and agglomeration of hard coal
1430	Mining of chemical and fertilizer minerals
1597	Manufacture of malt
1711	Preparation and spinning of cotton-type fibres
1810	Manufacture of leather clothes
2310	Manufacture of coke oven products
2413	Manufacture of other inorganic basic chemicals
2414	Manufacture of other organic basic chemicals
2415	Manufacture of fertilizers and nitrogen compounds
2417	Manufacture of synthetic rubber in primary forms
2710	Manufacture of basic iron and steel and of ferro-alloys
2731	Cold drawing
2742	Aluminium production
2744	Copper production
2745	Other non-ferrous metal production
2931	Manufacture of agricultural tractors

1.2. Based on the quantitative criteria set out in paragraph 15 of Article 10a of Directive 2003/87/EC

NACE Code	Description

1562	Manufacture of starches and starch products
1583	Manufacture of sugar
1595	Manufacture of other non-distilled fermented beverages
1592	Production of ethyl alcohol from fermented materials
2112	Manufacture of paper and paperboard
2320	Manufacture of refined petroleum products
2611	Manufacture of flat glass
2613	Manufacture of hollow glass
2630	Manufacture of ceramic tiles and flags
2721	Manufacture of cast iron tubes
2743	Lead, zinc and tin production

1.3. Based on the quantitative criteria set out in point (a) of Article 10a(16) of Directive 2003/87/EC

NACE Code	Description
2651	Manufacture of cement
2652	Manufacture of lime

1.4. Based on the quantitative criteria set out in point (b) of Article 10a(16) of Directive 2003/87/EC

NACE Code	Description
1110	Extraction of crude petroleum and natural gas
1310	Mining of iron ores
1320	Mining of non-ferrous metal ores, except uranium and thorium ores
1411	Quarrying of ornamental and building stone
1422	Mining of clays and kaolin

1450	Other mining and quarrying n.e.c
1520	Processing and preserving of fish and fish products
1541	Manufacture of crude oils and fats
1591	Manufacture of distilled potable alcoholic beverages
1593	Manufacture of wines
1712	Preparation and spinning of woollen-type fibres
1713	Preparation and spinning of worsted-type fibres
1714	Preparation and spinning of flax-type fibres
1715	Throwing and preparation of silk, including from noils, and throwing and texturing of synthetic or artificial filament yarns
1716	Manufacture of sewing threads
1717	Preparation and spinning of other textile fibres
1721	Cotton-type weaving
1722	Woollen-type weaving
1723	Worsted-type weaving
1724	Silk-type weaving
1725	Other textile weaving
1740	Manufacture of made-up textile articles, except apparel
1751	Manufacture of carpets and rugs
1752	Manufacture of cordage, rope, twine and netting
1753	Manufacture of non-wovens and articles made from non-wovens, except apparel
1754	Manufacture of other textiles n.e.c.
1760	Manufacture of knitted and crocheted fabrics
1771	Manufacture of knitted and crocheted hosiery
1772	Manufacture of knitted and crocheted pullovers, cardigans and similar articles
1821	Manufacture of workwear
1822	Manufacture of other outerwear

1823	Manufacture of underwear
1824	Manufacture of other wearing apparel and accessories n.e.c.
1830	Dressing and dyeing of fur; manufacture of articles of fur
1910	Tanning and dressing of leather
1920	Manufacture of luggage, handbags and the like, saddlery and harness
1930	Manufacture of footwear
2010	Sawmilling and planing of wood; impregnation of wood
2052	Manufacture of articles of cork, straw and plaiting materials
2111	Manufacture of pulp
2124	Manufacture of wallpaper
2215	Other publishing
2330	Processing of nuclear fuel
2412	Manufacture of dyes and pigments
2420	Manufacture of pesticides and other agro-chemical products
2441	Manufacture of basic pharmaceutical products
2442	Manufacture of pharmaceutical preparations
2452	Manufacture of perfumes and toilet preparations
2463	Manufacture of essential oils
2464	Manufacture of photographic chemical material
2465	Manufacture of prepared unrecorded media
2466	Manufacture of other chemical products n.e.c.
2470	Manufacture of man-made fibres
2511	Manufacture of rubber tyres and tubes
2615	Manufacture and processing of other glass, including technical glassware
2621	Manufacture of ceramic household and ornamental articles
2622	Manufacture of ceramic sanitary fixtures

2623	Manufacture of ceramic insulators and insulating fittings
2624	Manufacture of other technical ceramic products
2625	Manufacture of other ceramic products
2626	Manufacture of refractory ceramic products
2681	Production of abrasive products
2722	Manufacture of steel tubes
2741	Precious metals production
2861	Manufacture of cutlery
2862	Manufacture of tools
2874	Manufacture of fasteners, screw machine products, chain and springs
2875	Manufacture of other fabricated metal products n.e.c.
2911	Manufacture of engines and turbines, except aircraft, vehicle and cycle engines
2912	Manufacture of pumps and compressors
2913	Manufacture of taps and valves
2914	Manufacture of bearings, gears, gearing and driving elements
2921	Manufacture of furnaces and furnace burners
2923	Manufacture of non-domestic cooling and ventilation equipment
2924	Manufacture of other general purpose machinery n.e.c.
2932	Manufacture of other agricultural and forestry machinery
2941	Manufacture of portable hand held power tools
2942	Manufacture of other metalworking machine tools
2943	Manufacture of other machine tools n.e.c.
2951	Manufacture of machinery for metallurgy
2952	Manufacture of machinery for mining, quarrying and construction
2953	Manufacture of machinery for food, beverage and tobacco processing
2954	Manufacture of machinery for textile, apparel and leather production

2955	Manufacture of machinery for paper and paperboard production
2956	Manufacture of other special purpose machinery n.e.c.
2960	Manufacture of weapons and ammunition
2971	Manufacture of electric domestic appliances
3001	Manufacture of office machinery
3002	Manufacture of computers and other information processing equipment
3110	Manufacture of electric motors, generators and transformers
3120	Manufacture of electricity distribution and control apparatus
3130	Manufacture of insulated wire and cable
3140	Manufacture of accumulators, primary cells and primary batteries
3150	Manufacture of lighting equipment and electric lamps
3162	Manufacture of other electrical equipment n.e.c.
3210	Manufacture of electronic valves and tubes and other electronic components
3220	Manufacture of television and radio transmitters and apparatus for line telephony and line telegraphy
3230	Manufacture of television and radio receivers, sound or video recording or reproducing apparatus and associated goods
3310	Manufacture of medical and surgical equipment and orthopaedic appliances
3320	Manufacture of instruments and appliances for measuring, checking, testing, navigating and other purposes, except industrial process control equipment
3340	Manufacture of optical instruments and photographic equipment
3350	Manufacture of watches and clocks
3511	Building and repairing of ships
3512	Building and repairing of pleasure and sporting boats
3530	Manufacture of aircraft and spacecraft
3541	Manufacture of motorcycles
3542	Manufacture of bicycles
3543	Manufacture of invalid carriages
3550	Manufacture of other transport equipment n.e.c.

3621	Striking of coins
3622	Manufacture of jewellery and related articles n.e.c.
3630	Manufacture of musical instruments
3640	Manufacture of sports goods
3650	Manufacture of games and toys
3661	Manufacture of imitation jewellery
3662	Manufacture of brooms and brushes
3663	Other manufacturing n.e.c.

2. Beyond NACE-4 level based on the quantitative criteria set out in paragraphs 15 and 16 of Article 10a of Directive 2003/87/EC

Prodcom Code	Description
15331427	Concentrated tomato puree and paste
155120	Milk and cream in solid forms
155153	Casein
155154	Lactose and lactose syrup
15891333	Dry bakers' yeast
24111150	Hydrogen (including the production of hydrogen in combination with syngas).
24111160	Nitrogen
24111170	Oxygen
243021	Prepared pigments, opacifiers and colours, vitrifiable enamels and glazes, engobes, liquid lustres and the like; glass frit
24621030	Gelatin and its derivatives; isinglass (excluding casein glues and bone glues)
261411	Slivers, rovings, yarn and chopped strands, of glass fibre
26821400	Artificial graphite, colloidal, semi-colloidal graphite and preparations
26821620	Exfoliated vermiculite, expanded clays, foamed slag and similar expanded mineral materials and mixtures thereof

3. At NACE-4 level based on the qualitative criteria set out in paragraph 17 of Article 10a of Directive 2003/87/EC

NACE Code	Description
1730	Finishing of textiles
2020	Manufacture of veneer sheets; manufacture of plywood, laminboard, particle board, fibre board and other panels and boards
2416	Manufacture of plastics in primary forms
2751	Casting of iron
2753	Casting of light metals

Source: European Commission, *Commission Decision of 24 December 2009 determining, pursuant to Directive 2003 /8 7/EC of the European Parliament and of the Council, a list of sectors and subsectors which are deemed to be exposed to a significant risk of carbon leakage* (Brussels: 2009).

End Notes

[1] For a further discussion, see CRS Report R40 100, *"Carbon Leakage" and Trade: Issues and Approaches*, by Larry Parker and John Blodgett.

[2] Department of Climate Change, Commonwealth of Australia, *Carbon Pollution Reduction Scheme: Green Paper* (July 2008), p. 292.

[3] For more information on trade and carbon leakage issues, see CRS Report R40 100, *"Carbon Leakage" and Trade: Issues and Approaches*, by Larry Parker and John Blodgett.

[4] For further background on Phase 3 of the ETS, see CRS Report R41049, *Climate Change and the EU Emissions Trading Scheme (ETS): Looking to 2020*, by Larry Parker

[5] For further background on the ETS, see CRS Report RL34 150, *Climate Change and the EU Emissions Trading Scheme (ETS): Kyoto and Beyond*, by Larry Parker.

[6] More information, including relevant directives, on the EU-ETS is available on the European Union's website at http ://europa. eu.int/scadplus/leg/en/lvb/l280 12.htm.

[7] Directive 2003/87/EC of the European Parliament and of the Council of 13 October 2003 establishing a scheme for greenhouse gas emissions allowance trading within the Community and amending Council Directive 96/61/EC.

[8] Commission of the European Communities, *Communication from the Commission: Progress towards Achieving the Kyoto Objectives* (November 19, 2008).

[9] A. Denny Ellerman and Barbara K. Buchner, "The European Union Emissions Trading Scheme: Origins, Allocations, and Early Results," *Environmental Economics and Policy* (Winter 2007), pp. 69-70; and International Emissions Trading Association, "IETA Position Paper on EU ETS Marking Functioning," (no date), p. 3.

[10] For more on the effect of the ETS on Kyoto mechanisms, see A. Denny Ellerman and Barbara K. Buchner, "The European Union Emissions Trading Scheme: Origins, Allocations, and Early Results," *Environmental Economics and Policy* (Winter 2007), p. 84; and International Emissions Trading Association, "IETA Position Paper on EU ETS Market Functioning" (no date), p. 2. For more information on the Kyoto Protocol mechanisms, see CRS Report RL33826, *Climate Change: The Kyoto Protocol, Bali "Action Plan," and International Actions*, by Jane A. Leggett.

[11] A Denny Ellerman and Barbara K. Buchner, "The European Union Emissions Trading Scheme: Origins, Allocations, and Early Results," *Environmental Economics and Policy* (Winter 2007), p. 84.

[12] European Commission, Directorate General for Environment, *Review of EU Emissions Trading Scheme: Survey Highlights*, (November 2005), pp. 5-7.

[13] As published in Carbon Trust, *EU ETS Impacts on Profitability and Trade* (January 2008), p. 3.

[14] The Value at Stake can be defined as the difference in costs between "business as usual" and reduced emissions scenarios (based both on the impact of increasing energy process and the potential for reducing consumption). This calculation includes both direct and indirect costs.

[15] See European Commission, *Directive 2009/29/EC of the European and of the Council of 23 April 2009 amending Directive 2003/87/EC so as to improve and extend the greenhouse gas emission allowance trading system of the Community* (Brussels, April 23, 2009).

[16] European Commission, *Directive 2009/29/EC of the European and of the Council of 23 April 2009 amending Directive 2003/87/EC so as to improve and extend the greenhouse gas emission allowance trading system of the Community* (Brussels, April 23, 2009), annex I.

[17] European Commission, *Directive 2009/29/EC of the European and of the Council of 23 April 2009 amending Directive 2003/8 7/EC so as to improve and extend the greenhouse gas emission allowance trading system of the Community* (Brussels, April 23, 2009), paragraph 21.

[18] European Commission, *Commission Decision of 24 December 2009 determining, pursuant to Directive 2003/87/EC of the European Parliament and of the Council, a list of sectors and subsectors which are deemed to be exposed to a significant risk of carbon leakage* (Brussels, 2009).

[19] European Commission, *Draft Commission Decision of 18 September 2009 determining, pursuant to Directive 2003/87/EC of the European Parliament and of the Council, a list of sectors and subsectors which are deemed to be exposed to a significant risk of carbon leakage* (Brussels, 2009).

[20] European Commission, *Directive 2009/29/EC of the European and of the Council of 23 April 2009 amending Directive 2003/87/EC so as to improve and extend the greenhouse gas emission allowance trading system of the Community* (Brussels, April 23, 2009), paragraph 27.

[21] For a further discussion of auctioning and the ETS, see Cameron Hepburn et. al., "Auctioning of EU ETS phase II allowances: how and why?" *Climate Policy* (2006), pp. 137-160.

[22] A Denny Ellerman and Barbara K. Buchner, "The European Union Emissions Trading Scheme: Origins, Allocations, and Early Results," *Environmental Economics and Policy* (Winter 2007), p. 73.

[23] Joseph Kruger, Wallace E. Oates, and William A. Pizer, "Decentralization in the EU Emissions Trading Scheme and Lessons for Global Policy," *Environmental Economics and Policy* (Winter 2007), p. 114.

[24] E.g., Deutsche Bank Research, *EU Emission Trading: Allocation Battles Intensifying* (March 6, 2007), pp. 2-3; and Regina Betz and Misato Sato, "Emissions Trading: Lessons Learnt from the 1st Phase of the EU ETS and Prospects for the 2nd Phase," 6 *Climate Policy* (2006), p. 353.

[25] A Denny Ellerman and Barbara K. Buchner, "The European Union Emissions Trading Scheme: Origins, Allocations, and Early Results," *Environmental Economics and Policy* (Winter 2007), p. 85.

[26] For a review of proposed NAP 2 auction proposals as of January 12, 2007, see Karsten Neuhoff, *EU ETS Auction Workshop*, (Cambridge, January 12, 2007), p. 26. NAP refers to the National Allocation Plans member countries submitted to the European Commission during Phase 1 and Phase 2 to demonstrate how they were going to meet their emissions target.

[27] Martina Priebe, *Distributional Effect of Carbon-Allowance Trading* (Cambridge, January 12, 2007). Also, see Eurochambres, *Review of the EU Emission Trading System* (June 2007), p. 5.

[28] Julia Reinaud, *Issues Behind Competitiveness and Carbon Leakage: Focus on Heavy Industry* (October 2008), p. 6.

[29] Carbon Trust, *EU ETS Impacts on Profitability and Trade* (January 2008), p. 4; and The Climate Group, *The Effects of EU Climate Legislation on Business Competitiveness; A Survey and Analysis* (September 2009), p. 8.

[30] Point Carbon, *Carbon 2009* (2009), p. 10.

[31] The Climate Group, *The Effects of EU Climate Legislation on Business Competitiveness; A Survey and Analysis,* (September 2009).

[32] See CRS Report R40100, *"Carbon Leakage" and Trade: Issues and Approaches*, by Larry Parker and John Blodgett, and Carbon Trust, *The European Emissions Trading Scheme: Implications for Industrial Competitiveness* (June 2004), pp. 6-7.

[33] Carbon Trust, *EU ETS Impacts on Profitability and Trade: A sector by sector analysis* (2008), p. 29.

[34] For more information on trade and carbon leakage issues, see CRS Report R40 100, *"Carbon Leakage" and Trade: Issues and Approaches*, by Larry Parker and John Blodgett.

[35] H.R. 2454, as passed, new section 761(a).

[36] H.R. 2454, as passed, new section 761(b).

[37] H.R. 2454, as passed, new section 761(c).

[38] H.R. 2454, as passed, new section 763(b).

[39] The provision also provides that iron and steel made with different processes and metal, soda ash, or phosphate production classified under more than one NAICS code be treated as different categories under the section; and that differences in use of combined heat and power technologies be taken into account.

[40] H.R. 2454, as passed, new section 763(b)(2)(C).

[41] The NAICS codes are updated every five years. H.R. 2454 specifically defines NAICS codes as 2002 NAICS codes. See 2002 NAICS Definition, 31-33, Manufacturing, at http://www.census.gov/cgi-bin/sssd/naics/naicsrch?chart_code= 3 1&search=2002%20NAICS%20Search.

[42] Testimony of John McMackin for the Energy Intensive Manufacturers' Working Group on Greenhouse Gas Regulation before the House Committee on Ways and Means (March 24, 2009), available at http://waysandmeans.house.gov/media

[43] The second set of criteria set up under Subpart 1 only adds one sector, lime manufacturing, to the list of presumptively covered sectors.

[44] U.S. Environmental Protection Agency, *Comparison of FTI and EPA analyses of H.R. 2454, Title IV*, Memorandum to the House Energy and Commerce Committee Staff (June 10, 2009).

[45] The data for NAICS 212210 and 212234 are for 2007.

[46] This calculation mixes two data sets and, therefore, should be viewed as a ballpark estimate. The data on purchases of electricity by the eligible sectors is from the U.S. Census Survey for 2006. The data for all retail sales is from the Energy Information Administration (EIA) for 2006. Based on Census data, the eligible industries purchased between 295 (EPA's List) and 315 (Working Group list) billion Kwhs in 2006. Based on EIA data, that is equal to about 29%- 31% of industrial sector retail sales or between 8% and 8.6% of total retail sales in 2006. All else being equal, eligible energy-intensive industry should get about 2.4% to 2.6% of total available allowances via the LDC pass-through (30% LDC allocation times 8%-8.6%).

[47] Email correspondence from Jared Creason, Ph.D., Climate Economics Branch, U.S. Environmental Protection Agency (September 30, 2009).

[48] Frederick F. Butler, President, National Association of Regulatory Utility Commissioners, Letter to Chairman Boxer and Ranking Member Inhofe (October 28, 2009), p. 2.

[49] U.S. Environmental Protection Agency, *EPA Analysis of the American Clean Energy and Security Act of 2009: H.R. 2454 in the 111th Congress—Appendix* (June 23, 2009), p. 42.

[50] EIA, *Energy Market and Economic Impacts of H.R. 2454, the American Clean Energy and Security Act of 2009* (August 2009), p. 44.

[51] More specifically, and as discussed later in this report, beginning June 30, 2018, and every four years thereafter, the President would be required to determine for each eligible industrial sector whether more than 85% of U.S. imports for that sector is from countries that are either (1) parties to international agreements requiring economy-wide binding national commitments at least as stringent as those of the United States, (2) have annual energy or greenhouse gas intensities for the sector comparable or better than the equivalent U.S. sector, or (3) parties to an international or bilateral emission reduction agreement for that sector. If not, the President would be required, no later than June 30, 2018 (and every four years thereafter), to assess the effectiveness of subpart 1 rebates and the international reserve allowance program in mitigating or potentially mitigating the carbon leakage in that sector, and respond by (1) modifying the rebate formula under subpart 1, and (2) implementing (or continuing to implement) an international reserve allowance program with respect to imports of covered goods from that sector.

[52] Also, under subpart 2, iron and steel produced by different processes shall be considered as one eligible industrial sector (Section 769). In contrast, subpart 1 would consider entities using different iron and steelmaking processes to be in different industrial sectors (Section 764(d)).

[53] H.R. 2454, as passed, new section 767(a).

[54] H.R. 2454, as passed, new section 767(b).

[55] Any such joint resolution would be considered under an expedited legislative procedure set out in section 152 of the Trade Act of 1974, 19 U.S.C. § 2191, providing for automatic discharge of the resolution from committee, a prohibition on amendments, and limited floor debate in the House and Senate. H.R. 2454, as passed, new section 767 (b)(2)-(3).

[56] H.R. 2454, as passed, new section 767(b)(1)(emphasis added). Although H.R. 2454 anticipates Senate or congressional approval of a multilateral GHG-reduction agreement, it does not appear to indicate how a sectoral agreement should be treated were such an agreement to be signed by the United States.

[57] Note that, for purposes of determining whether an industrial sector is eligible for free allowances, EPA would determine trade intensity by reference to the value of imports and exports (new section 763 (b)(2)(A)(iii)). See also the definition of "output" at H.R. 2454, as passed, new section 762(7) ("The term 'output' means the total tonnage or other standard unit of production (as determined by the Administrator) produced by an entity in an industrial sector).

[58] H.R. 2454, as passed, new section 767(d).

[59] H.R. 2454, as passed, new section 767(d)(2).

[60] H.R. 2454, as passed, new section 768(e).

[61] H.R. 2454, as passed, new section 768(a)(1)(A)-(D), (F)-(G).

[62] H.R. 2454, as passed, new section 768(b).

[63] H.R. 2454, as passed, new section 768(a)(1)(E).

[64] H.R. 2454, as passed, new section 768(a)(2). As noted earlier, H.R. 2454 also sets out as one of the purposes of subpart 2, "to ensure that the measures described in subpart 2 are designed and implemented in a manner

consistent with applicable international agreements to which the United States is a party." H.R. 2454, as passed, new section 761(c)(2).

[65] Center for Clean Air Policy, *Summary of Provisions to Protect the Competitiveness of U.S. Industry in the American Clean Energy and Security Act* (July 23, 2009), p. 24.

[66] EPA, *EPA Analysis of the Lieberman-Warner Climate Security Act of 2008:* S. 2191 *in 110th Congress* (March 14, 2008).

[67] EPA, *EPA Analysis of the Lieberman-Warner Climate Security Act of 2008:* S. 2191 *in 110th Congress* (March 14, 2008), p. 84.

[68] EPA, *EPA Analysis of the Lieberman-Warner Climate Security Act of 2008:* S. 2191 *in 110th Congress* (March 14, 2008), p. 85.

[69] The People's Republic of China, *Initial National Communication on Climate Change* (Beijing, October 2004).

[70] *Id.* at 4. Continuing on page 33 of the document, China says specifically with respect to the energy inventory: "Because existing statistical materials and data could not meet the needs for preparing the inventory, part of the activity data could only be obtained by adopting the methods of investigation and experts' judgment. For example, activity data by device in some important industries such as building material and metallurgy was based on experts judgment; owing to the lack of the measured data on emission factors from coal combustion by sector and by device, the relevant potential emission factors and oxidations rates could only be determined through case studies, questionnaires and partial supplementary measurements; due to the lack of detailed measurement data, methane emissions under different circumstances from different types of biomass stoves could only be estimated by using the same emissions factors. All those would affect the accuracy of energy inventory."

[71] United Nations Framework Convention on Climate Change, Subsidiary Body for Implementation, *Sixth compilation and Synthesis of Initial National Communications for Parties not included in Annex I to the Convention: Addendum: Inventories of anthropogenic emissions by sources and removals by sinks of greenhouse gases*, FCCC/SBI/2005/18/Add.2, (October 25, 2005), p. 10.

[72] National Bureau of Statistics of China, *China Statistical Yearbook- 2006*, p. 14-24, available at http://www.stats.gov.cn/tjsj/ndsj/2006/indexeh.htm.

[73] See the UNFCCC website at http://unfccc.int/di/DetailedByParty/Event.do?event=go.

[74] H.R. 2454, as passed, new section 762(2).

[75] The definition of "manufactured item for consumption" lists several requirements that must be met for a product to qualify as such, including that the good be one, "(ii) with respect to which an international reserve allowance program pursuant to Subpart 2 is in effect with regard to the eligible industrial sector and the quantity of international reserve allowances is not zero ..." While the definition would make item (ii) applicable to the good for which the petition is filed, it would appear that item (ii) could logically apply only to the goods that are contained in the manufactured item. First, the producers filing the petition to include the particular "manufactured item for consumption" as a "covered good" could not have been part of an eligible industrial sector with an IRA program since, if they were, their goods would already have been at least potentially subject to IRAs. Second, the good for which the petition is filed could not be one for which the quantity of IRAs is greater than zero, since the good would not yet be included in the sectoral program.

[76] The term "like product" is used in GATT obligations involving most-favored nation and national treatment, which prohibit WTO Members from discriminating between like products imported from different countries or between like imported and domestic items. "Likeness" is ordinarily determined by comparing products under four criteria: physical properties, end-uses, consumer preferences, and tariff classification. See, generally, World Trade Organization, WTO ANALYTICAL INDEX; GUIDE TO WTO LAW AND PRACTICE 145-48, 163-67 (2d ed. 2007).

[77] H.R. 2454. as passed, new section 765(a)-(b).

[78] H.R. 2454, as passed, new section 766(a).

[79] H.R. 2454, as passed, new section 766(b).

[80] The bill does not indicate how many countries other than the United States must be a party to the multilateral agreement for it to be acceptable to the Senate or the Congress under these provisions. By definition, a multilateral agreement qualifies as such so long as it has at least three parties.

[81] H.R. 2454, as passed, new section 767(b)(1).

[82] H.R. 2454, as passed, new section 767(b)(4).

[83] See Vienna Convention on the Law of Treaties art. 26 ("Every treaty in force is binding upon the parties to it and must be performed by them in good faith"); id. art. 27 ("A party may not invoke the provisions of its internal law as justification for its failure to perform a treaty"); and RESTATEMENT (THIRD) OF THE FOREIGN RELATIONS LAW OF THE UNITED STATES § 111, reporters' note h (1987).

Note, for example, the history of the multilateral Basel Convention on the Control of the Transboundary Movements of Hazardous Wastes and their Disposal. When the President submitted the Convention to the Senate in 1991, the accompanying transmittal notice stated that "[b]efore the United States can deposit its instrument of ratification, changes in domestic law will be needed." S. Treaty Doc. 102-5, at X (1991). While the executive branch proposed implementing legislation to the Congress at the time and the Senate gave its consent to the Convention in August 1992, 138 Cong. Rec. 22,861 (1992), implementing legislation has not yet been enacted and the United States has not become a Convention party. For additional background information on the deposit of instruments of ratification for a treaty or international agreement, see, generally, *Treaties and Other International Agreements: The Role of the United States Senate; A Study Prepared for the Senate Committee on Foreign Relations by the Congressional Research Service* 147- 50 (Jan. 2001)(S. Prt. 106-71).

Implementing legislation may also address other implementation issues, such as the relationship of the agreement to federal and state law and whether private rights of action based on the agreement are allowed. See, for example, Uruguay Round Agreements Act, P.L. 103-465, § 102, 19 U.S.C. § 3512.

[84] It is unclear whether the implementing legislation referenced in new section 767(b)(4), the provision setting out how consistency with U.S. negotiating objectives would be established for a congressional-executive agreement, would also include provisions implementing the agreement as a matter of domestic law. *Cf.* Trade Act of 2002, § 2103(b)(3)(B), 19 U.S.C. § 2103(b)(3)(B), distinguishing between legislative provisions approving a trade agreement and provisions making changes in domestic law to implement the agreement. Moreover, if the multilateral GHG-reduction agreement is approved as a treaty, separate implementing legislation may still be needed.

[85] The discussion here is restricted to provisions of H.R. 2454 and does not explicitly address the program developed by the EU, although there may be some similarities.

[86] The WTO-consistency of such measures, particularly border requirements, has been the subject of considerable legal commentary, including discussion in a 2009 report prepared jointly by the United Nations Environment Program (UNEP) and the World Trade Organization. See *Trade and Climate Change; A report by the United Nations Environment Programme and the World Trade Organization* 90-110 (2009), at http://www.wto.org/english/res_e/ booksp_e/trade_climate_change_e.pdf. A number of these commentaries are referenced in the UNEP/WTO report.

[87] The United States is also party to number of bilateral and regional free trade agreements (FTAs), including the North American Free Trade Agreement (NAFTA) and the Central America-Dominican Republic-United States Free Trade Agreement (DR-CAFTA), and bilateral agreements with such countries as Australia and Chile. These agreements incorporate certain GATT rights and obligations, such as national treatment of imported goods, a prohibition on quantitative restrictions, and general exceptions for measures that are inconsistent with agreement obligations, but also contain, among other things, their own tariff obligations, rules of origin, and dispute settlement procedures and, for the NAFTA, a chapter on energy trade. While the requirements of these agreements are not addressed in this report, it is important to note that two of the largest U.S. trading partners, Canada and Mexico, are parties to the NAFTA. To the extent that these countries are exporters of the types of products that are likely to be produced by eligible industrial sectors or later determined to be remediable "manufactured items for consumption," relevant NAFTA obligations may also need to be considered by EPA in applying an IRA program to Canadian or Mexican goods.

[88] The General Agreement on Tariffs and Trade 1994 (GATT 1994), which consists of the GATT, as originally adopted in 1947 (GATT 1947) as well as subsequent GATT decisions, waivers, and other provisions, maybe accessed at http://www.wto.org/english/docs_e/legal_e/06-gatt.pdf, and http://www.wto.org/english/docs_e/legal_e/gatt47_e.pdf.

[89] See, for example, *Elements of a Trade and Climate Code,* in Gary Clyde Hufbauer, Steve Charnovitz, and Jisun Kim, GLOBAL WARMING AND THE WORLD TRADING SYSTEM 103-110 (2009). Note also the now-expired Article 8.2(c) of the WTO Agreement on Subsidies and Countervailing Measures (SCM Agreement) exempting from WTO challenge certain "assistance to promote adaptation of existing facilities to new environmental requirements imposed by law and/or regulations which result in greater constraints and financial burden on firms," so long as the assistance did not constitute a prohibited subsidy, in other words, an export subsidy or a subsidy contingent on the use of domestic over imported products. Assistance was exempted under this provision even though it may have been specific to an industry or group of industries, a condition that would ordinarily make a non-prohibited subsidy actionable. The SCM Agreement may be accessed at http://www.wto.org/english/docs_e/legal_e/24-scm.pdf.

[90] See, for example, WTO Agreement on Agriculture art. 13 (making certain domestic agricultural support and agricultural export subsidies exempt from WTO dispute settlement actions for the initial nine years of the Agreement); Agreement on Trade-Related Aspects of Intellectual Property Rights art. 64.2 (exempting non-violation claims—claims based only on trade injury and not on violations of the Agreement—from WTO dispute settlement for the initial five years of the Agreement; moratorium since extended). The Agreement on Agriculture may be accessed at http://www.wto.org/english/docs_e/legal_e/14-ag.pdf; the Agreement on Trade-Related Aspects of Intellectual Property Rights (TRIPS) may be accessed at http://www.wto.org/english/docs_e/legal_e/27-trips.pdf.

[91] Note, for example, WTO, Kimberley Process Certification Scheme for Rough Diamonds; Decision of 15 May 2003, WT/L/518 (May 27, 2003) (WTO waiver through December 31, 2006, for certain actions taken by WTO Members to control diamond trade pursuant to international agreement); WTO, Kimberley Process Certification Scheme for Rough Diamonds; Decision of 15 December 2006, WT/L/676 (December 19, 2006)

(extension of waiver through December 31, 2012); WTO, Amendment of the TRIPS Agreement; Decision of 6 December 2005, WT/L/641 (December 8, 2005) (amendment of Agreement on Trade-Related Intellectual Property Rights to permit exporting Members to require compulsory licensing for the production of pharmaceutical products and to export such products to eligible importing Members). See also *WTO Members Agree to Further Extension of TRIPS/Medicines Ratification Deadline*, 26 Int'l Trade Rep. (BNA) 1497 (November 5, 2009).

[92] See, for example, Pascal Lamy, Director-General, WTO, Climate First, Trade Second—GATTzilla is Long Gone, Address at Carleton University, Ottawa, Canada (November 2, 2009), at http://www.wto. org/english/ news_e/sppl_e/ sppl140_e.htm.

[93] In the event that not all WTO Members were party to a WTO-related segment of a multilateral GHG agreement or to a separate international agreement providing for measures to address competitiveness and leakage issues, the legal situation of non-party WTO Members would need to be addressed.

[94] For additional information on WTO dispute settlement, see CRS Report RS20088, *Dispute Settlement in the World Trade Organization (WTO): An Overview*, by Jeanne J. Grimmett, and the WTO website at http://www.wto.org/ english/tratop_e/dispu_e/dispu_e.htm. The WTO Dispute Settlement Understanding may be accessed at http://www.wto.org/english/docs_e/legal_e/28-dsu.pdf.

[95] Under U.S. law, a WTO decision finding that a federal law is inconsistent with a WTO obligation cannot be implemented unless Congress amends or repeals the statute, as the case may be. WTO decisions faulting a U.S. agency regulation or practice may be implemented through administrative action under existing authorities, provided that procedures set out in § 123(g) of the Uruguay Round Agreements Act, 19 U.S.C. § 3533(g), are followed. For further discussion, see CRS Report RS22 154, *World Trade Organization (WTO) Decisions and Their Effect in U.S. Law*, by Jeanne J. Grimmett, and CRS Report RL32014, *WTO Dispute Settlement: Status of U.S. Compliance in Pending Cases*, by Jeanne J. Grimmett.

[96] A disputing party may also request that a compliance panel be established to determine whether the defending Member has complied with a WTO ruling.

[97] Appellate Body Report, *United States—Anti-dumping Act of 1916*, paras. 60-61, WT/DS136/AB/R, WT/DS162/AB/R (August 28, 2000).

[98] Appellate Body Report, *United States—Sunset Review of Anti-Dumping Measures on Oil Country Tubular Goods from Argentina*, para. 172, WT/DS268/AB/R (November 29, 2004). The Appellate Body further described "as such" claims as follows: "By definition, an 'as such' claim challenges laws, regulations, or other instruments of a Member that have general and prospective application, asserting that a Member's conduct—not only in a particular instance that has occurred, but in future situation as well—will necessarily be inconsistent with that Member's WTO obligations. In essence, complaining parties bringing 'as such' challenges seek to prevent Members *ex ante* from engaging in certain conduct. The implications of such challenges are obviously more far-reaching than 'as applied' claims." *Id.*

[99] See cases cited in Panel Report, *United States—Laws, Regulations and Methodology for Calculating Dumping Margins ("Zeroing")*, para. 7.55, n. 158, WT/DS294.R (October 31, 2005).

[100] Appellate Body Report, *United States—Sunset Review of Anti-Dumping Duties on Corrosion-Resistant Carbon Steel Flat Products from Japan*, para. 93, WT/DS244/AB/R (December 15, 2003). Note also Panel Report, *United States— Sections 301-310 of the Trade Act of 1974*, WTO/DS152/R (December 22, 1999) (legislation granting discretionary powers may be found to be inconsistent "as such" if it does not create a strong legal basis for WTO-consistent action).

[101] An "allowance" is defined in H.R. 2454, as passed, as "a limited authorization to emit, or have attributable greenhouse gas emissions in an amount of, 1 ton of carbon dioxide equivalent of a greenhouse gas in accordance with this title. Such term includes an emission allowance ..." H.R. 2454, section 321, as passed, adding new section 700(5).

[102] Letter of Douglas W. Elmendorf, Director, Congressional Budget Office, to Hon. Henry A. Waxman, Chairman, House Committee on Energy and Commerce, May 15, 2009, at [1], at http://www.cbo. gov/ftpdocs/ 102xx/doc10232/5- 15-WaxmanLetter.pdf, (hereinafter, CBO letter).

[103] *Id.* at [1]-2.

[104] *Id.* at 2.

[105] Joint Committee on Taxation, *Climate Change Legislation: Tax Considerations; Scheduled for a Public Hearing Before the Senate Committee on Finance on June 16, 2009*, at 5-6 (June 12, 2009), at http://www.jct.gov/ publications.html?func=startdown&id=3559.

[106] *Id.* at 6.

[107] *Id.*

[108] *Id.*

[109] *Id.* at 7-11

[110] GATT 1994 art. VI:5:6(a). The GATT 1994 and the SCM Agreement define a countervailing duty as "a special duty levied for the purpose of offsetting any subsidy bestowed directly or indirectly upon the manufacture, production or export of any merchandise." GATT art. VI:3; SCM Agreement art. 10, n.36.

[111] WTO panels have not yet addressed whether, or the extent to which, GATT Article XX exceptions may be linked and applied to obligations in other WTO agreements. See Panel Report, *China—Measures Affecting Trading Rights and Distribution Services for Certain Publications and Audiovisual Entertainment Products,* paras. 7.708-7.863, WT/DS363/R (August 12, 2009) (panel determined that China did not make case that measures found to be inconsistent with its WTO Accession Protocol fell within the scope of GATT Article XX(a) without panel first determining whether Article XX(a) applied to the Accession Protocol); see also Notification of an Appeal by China, *China—Measures Affecting Trading Rights and Distribution Services for Certain Publications and Audiovisual Entertainment Products,* at 2, WT/DS363/10 (September 23, 2009).

[112] SCM Agreement art. 1. 1(a)(1), (a)(2).

[113] SCM Agreement art. 1.1(a)(1).

[114] See, generally, World Trade Organization, GUIDE TO GATT LAW AND PRACTICE; ANALYTICAL INDEX 445-48 (updated 6th ed. 1995).

[115] Appellate Body Report, *Canada—Measures Affecting the Export of Civilian Aircraft,* paras. 149-157, WT/DS70/AB/R (August 2, 1999) (hereinafter, Canada Aircraft AB Report).

[116] SCM Agreement arts. 1.2.

[117] SCM Agreement art. 2.3.

[118] SCM Agreement arts. 21.(a), (c).

[119] SCM Agreement art. 2. 1(b)(footnote omitted).

[120] SCM Agreement. art. 2.1(b), n.2.

[121] SCM Agreement art. 3.1(a), (b). The WTO Appellate Body has determined that import substitution subsidies may be contingent in law or "in fact," notwithstanding that the prohibition does not contain the quoted language. Appellate Body Report, *Canada—Certain Measures Affecting the Automotive Industry,* paras. 137-43, WT/DS139/AB/R, WT/DS142/AB/R (May 31, 2000) (hereinafter, Canada Autos AB Report).

[122] SCM Agreement art 3.1(a), n.4.

[123] SCM Agreement art. 5.

[124] See Articles 6.4 and 6.5 of the SCM Agreement for further explanation of the terms used in Article 6.3.

[125] The SCM Agreement states that, for purposes of the Agreement, the term "like product" means "a product which is identical, i.e. alike in all respects to the product under consideration [i.e. the subsidized product], or in the absence of such a product, another product which, although not alike in all respects, has characteristics closely resembling those of the product under consideration." SCM Agreement art. 15.1, n.46.

[126] SCM Agreement art. 7.8.

[127] SCM Agreement art. 7.9.

[128] SCM agreement, art. 7.4, n.20.

[129] As discussed earlier, CBO has stated that, for budgetary purposes, distributed allowances should be recorded as outlays and the creation of emission allowances as revenues. CBO Letter, *supra* note 98, at [1]. Note also that the WTO Agreement on Agriculture treats "budgetary outlays" as subsidies for purposes of calculating a WTO Member's aggregate domestic support for agricultural products, Agreement on Agriculture annex 3, para. 2, and that the SCM Agreement includes, as the last item in its Illustrative List of Export Subsidies, "*[a]ny other charge on the public account* constituting an export subsidy in the sense of Article XVI of GATT 1994." SCM Agreement annex I, para. (*l*) (emphasis added). Discussion of domestic budgetary treatment of emission allowances for purposes of the WTO subsidy definition is beyond the scope of this report.

[130] The role of private payments in a subsidy scheme was addressed in *Canada— Measures Affecting the Importation of Milk and the Exportation of Dairy Products* (WT/DS 103), where the WTO Appellate Body upheld a WTO panel finding that producer-financed payments to support the export of dairy products were covered by commitments to reduce export subsidies contained in WTO Agreement on Agriculture. Article 9.1(c) of the Agreement provides that reduction commitments apply to listed export subsidies, including "payments on the export of an agricultural product that are financed by virtue of governmental action, whether or not a charge on the public account is involved...." The Appellate Body first upheld the panel's finding that the provision of milk at discounted prices to processors for export under the challenged program constituted payments, though in a form other than money, within the meaning of the Article 9.1(c). Appellate Body Report, *Canada—Measures Affecting the Importation of Milk and the Exportation of Dairy Products,* paras. 113, WT/DS103/AB/R, WTO/DS1 13/AB/R (October 13, 1999). The Appellate Body then upheld the panel's finding that producer-financed payments fell within the scope of the Article, provided they were "financed by virtue of governmental action." In upholding the panel, the Appellate Body stated that it was appropriate to look at governmental action as a whole in the payment system at issue and found that although the "'cost of selling milk at a reduced price for export is not borne by the government', 'governmental action' is, in our view, indispensible to the transfer of resources that take place at as a result of the operation" of the program. *Id.* paras. 119-120. The Appellate Body found that governmental action was involved at every stage of the program and that, in the regulatory framework involved, "'government agencies' stand so completely between the producers of the milk and the processors or the exporter that we have not doubt that the transfer of resources takes place, by virtue of governmental action." *Id.* para. 120. While the SCM Agreement does not contain language stating that subsidies include "payments ... that are financed by virtue of governmental

action, whether or not a charge on the public account is involved," the existence of such language in another WTO agreement addressing subsidization may indicate that WTO Members generally contemplate that the level of governmental involvement in the actual realization of wealth by the beneficiary of a government program would exceed the mere allocation of an economic instrument to that beneficiary. Note also Panel Report, *Japan—Measures Affecting Consumer Photographic Film and Paper*, para. 10.49, WT/DS44/R (March 31, 1998) ("non-binding [governmental] actions, which include sufficient incentives or disincentives for private parties to act in a particular manner, can potentially have adverse effects on competitive conditions of market access" so as to nullify or impair legitimately expected WTO benefits for purposes of filing a non-violation complaint—a challenge to a non-violative measure of another Member on the grounds that it causes WTO-related trade injury).

[131] Appellate Body Report, *Countervailing Duty Determination with Respect to Certain Softwood Lumber from Canada*, para. 75, WT/DS257/AB/R (January 19, 2004).

[132] See, for example, Javier de Cendra, *Can Emissions Trading Schemes be Coupled with Border Tax Adjustment? An Analysis vis à vis WTO Law*, 15 RECIEL 131, 137 (2006); Jacob Werksman, *Greenhouse Gas Emissions and the WTO*, 8 RECIEL 251, 255 (1999).

[133] Note, for example, the definition of a "commodity" in IRS Regulation section 1.954.1(f)(2)(i) as "tangible personal property of a kind that is actively traded or with respect to which the contractual interests are actively traded." See, generally, Matthew P. Haskins, *Green Trading in Carbon Emission Rights,* 122 TAX NOTES 387-88 (2009).

[134] World Trade Organization, WTO ANALYTICAL INDEX; GUIDE TO WTO LAW AND PRACTICE 755 (2d ed. 2007) (hereinafter, WTO ANALYTICAL INDEX).

[135] *Id.* at 754-57.

[136] See Canada Autos AB Report, *supra* note 117, paras. 87-94 (import duty exemption treated as foregoing of revenue that is otherwise due).

[137] In determining whether a benefit exists in cases involving the granting of loans or loan guarantees or the provision of goods or services, panels will examine whether the terms placed on the transaction by the government are more favorable than those available to the recipient in the marketplace. See, generally, WTO ANALYTICAL INDEX, *supra* note 130, at 761-73. See also Article 14 of the SCM Agreement, which sets out rules for calculating the benefit to a recipient where the governmental financial contribution consists of a equity capital, a loan, a loan guarantee, the provision of goods or services, or the purchase of goods.

[138] Panel Report, *United States—Tax Treatment for "Foreign Sales Corporations,"* para 7.103, WT/DS108/R (October 8, 1999).

[139] Panel Report, *United States—Tax Treatment for "Foreign Sales Corporations"; Recourse to Article 21.5 of the DSU by the European Communities,* para 8.46, WT/DS108/RW (August 20, 2001). As the United States had focused its arguments on the existence, or not, of a financial contribution, it again did not contest that a benefit would be conferred were the requisite financial contribution to be found. *Id.* para 8.47.

[140] Panel Report, *Canada—Certain Measures Affecting the Automotive Industry*, para. 10.165, WT/DS139/AB/R, WT/DS142/AB/R (February 11, 2000).

[141] Canada Aircraft AB Report, *supra* note 111, paras. 153-156.

[142] Appellate Body Report, *Japan—Taxes on Alcoholic Beverages*, at 23, WT/DS8/AB/R, WTDS10/AB/R, WT/DS1 1/AB/R (October 4, 1996) (hereinafter, Japan Beverages AB Report).

[143] Panel Report, *Argentina— Measures Affecting the Export of Bovine Hides and the Import of Finished Leather*, para. 11.182, WT/DS155/R (December 19, 2000).

[144] See, for example, Panel Report, *United States—Taxes on Petroleum and Certain Imported Substances,* L/6 175 (June 17, 1987), GATT, B.I.S.D. (34th Supp.) at 136 (1988); Japan Beverages AB Report, *supra* note 138, at 16.

[145] Appellate Body Report, *European Communities—Measures Affecting Asbestos and Asbestos-Containing Products*, para. 101 WT/DS135/AB/R (March 12, 2001) (hereinafter, EC Asbestos AB Report). "Like product" determinations under Article III are made on a case-by-case basis under a significant body of GATT/WTO jurisprudence. See, generally, *id.* at paras. 87-103 and WTO ANALYTICAL INDEX, *supra* note 130, at 145-48, 163-67.

[146] See Panel Report, *United States—Import Prohibition of Certain Shrimp and Shrimp Products,* paras. 7.11-7.17, WT/DS58/R (May 15, 1998). Although there is language in WTO decisions indicating that less favorable treatment of a like imported product may be permitted if it can be explained by factors unrelated to foreign origin (e.g., Appellate Body Report, *Dominican Republic—Measures Affecting the Importation and Internal Sale of Cigarettes,* para. 96, WT/DS302/AB/R (April 25, 2005)), there has not yet been a WTO case in which a panel or the Appellate Body has ratified less favorable treatment of an imported good under GATT Article III based on the fact that it was produced by means of a particular non-product-related processing or production method (PPM).

[147] Panel Report, *India—Measures Affecting the Automotive Sector*, para. 7.224, WT/DS146/R, WT/DS175/R (December 21, 2001).

[148] Appellate Body Report, *United States—Import Prohibition of Certain Shrimp and Shrimp Products; Recourse to Article 21.5 of the DSU by Malaysia,* paras.136-38,WT/DS58/AB/RW (October 22, 2001) (hereinafter, U.S. Shrimp (Article 21.5) AB Report).

[149] For further discussion, see CRS Report R40100, *"Carbon Leakage" and Trade: Issues and Approaches*, by Larry Parker and John Blodgett.

[150] Report of the Panel, *European Community Programme of Minimum Import Prices, Licenses and Surety Deposits for Certain Processed Fruits and Vegetables*, para. 4.15, L/4687 (adopted October 18, 1978), at http://www.wto.org/ gatt_docs/English/SULPDF/90950205.pdf. For the Article II:2(a) exception for internal taxes to apply, it would have to be shown that the import surcharge were "equivalent" to an Article III:2 internal tax applied "in respect of the like domestic product or in respect of a product from which the imported product has been produced in whole or in part."

[151] Panel Report, *United States—Import Measures on Certain Products from the European Communities*, paras. 6.62- 6.67, WT/DS165/R (July 17, 2000).

[152] Appellate Body Report, *United States—Import Measures on Certain Products from the European Communities*, para. 100, WT/DS165/AB/R (July 17, 2000). The Appellate Body stated, however, that it agreed with the apparent U.S. concession during oral argument that, in light of the panel finding on charges and costs, the increased bonding requirements themselves were inconsistent with this portion of Article II:1(b). *Id.* If challenged, the IRA requirements, aside from the fees paid by individual importers, may similarly be found to be inconsistent with the cited obligation.

[153] In addition, some GATT-inconsistent measures may qualify for the more generic Article XX(d) exception for "measures necessary to secure compliance with laws or regulations which are not inconsistent with" the GATT.

[154] See Appellate Body Report, *United States—Standards for Reformulated and Conventional Gasoline*, at 16, WT/DS2/AB/R (April 29, 1996) (hereinafter, U.S. Gasoline AB Report).

[155] In *European Communities—Tariff Preferences for Developing Countries*, the European Communities (EC) argued that special tariff preferences for countries combating drug production and trafficking could be justified under Article XX(b) on the ground that they were necessary to protect human life and health in the EC by supporting measures in drug-producing and trafficking countries that would reduce the supply of drugs into the EC. The panel was willing to examine whether EC health protection was a policy objective under the program , but was unable to find official evidence of this objective. Panel Report, *European Communities—Tariff Preferences for Developing Countries*, paras. 7.180-7.183 7.20 1, WT/DS246/R (December 1, 2003). In *United States—Import Prohibition of Certain Shrimp and Shrimp Products* (*U.S .Shrimp*), the Appellate Body specifically did not address whether there is an implied jurisdictional limitation in Article XX(g), but did find that, because the sea turtles which the challenged statute sought to protect were known to occur in waters under U.S. jurisdiction, there was a "sufficient nexus" between the endangered marine species at issue and the United States for purposes of the exception. Appellate Body Report, *United States—Import Prohibition of Certain Shrimp and Shrimp Products*, para. 133 (hereinafter, U.S. Shrimp AB Report).

[156] Appellate Body Report, *Brazil—Measures Affecting Imports of Retreaded Tyres*, para. 178 (December 3, 2007) (hereinafter, Brazil Tyres AB Report).

[157] *Id.*

[158] Appellate Body Report, *Korea—Measures Affecting Imports of Fresh, Chilled and Frozen Beef*, para. 162. WT/DS161/AB/R, WT/DS169/R (December 11, 2000); see also EC Asbestos AB Report, *supra* note 141, at para. 172.

[159] Brazil Tyres AB Report, *supra* note 152, paras. 150-151.

[160] *Id.* para. 151. The level of contribution of a measure can be demonstrated, for example, by evidence or data relating to the past or present or "quantitative projections in the future, or qualitative reasoning based on a set of hypotheses that are tested and supported by sufficient evidence." *Id.* See also U.S.-Gasoline AB Report, *supra* note 151, at 21.

[161] Brazil Tyres AB Report, *supra* note 152, para. 210.

[162] U.S. Shrimp AB Report, *supra* note 151, paras. 127-131.

[163] Id. at para. 128. See also *Panel Report, United States—Standards for Reformulated and Conventional Gasoline*, para. 6.37 WT/DS2.R (January 29, 1996) (a policy to reduce the depletion of clean air found to be a policy to conserve a natural resource for purposes of Article XX(g)).

[164] See *supra* note 151 and accompanying text.

[165] E.g., U.S. Gasoline AB Report, *supra* note 150, at 21.

[166] U.S. Shrimp AB Report, *supra* note 151, paras. 136-37.

[167] U.S. Gasoline AB Report, *supra* note 150, at 21.

[168] U.S. Shrimp AB Report, *supra* note 151, paras. 143-144.

[169] *Id.* paras. 158-59. In considering the U.S. restriction on shrimp caught with methods harmful to sea turtles, the Appellate Body stated that its analysis of the proviso would be colored by the preambular language to the WTO Agreement, which conditioned the trade objectives originally set out in the GATT with the following statement: "... while allowing for the optimal use of the world's resources in accordance with the objective of

sustainable development, seeking both to protect and preserve the environment and to enhance the means for doing so in a manner consistent with their respective needs and concerns at different levels of economic development."

[170] Brazil Tyres AB Report, *supra* note 152, para. 225.

[171] U.S. Gasoline AB Report, *supra* note 150, at 26-28.

[172] *Id.* at 28.

[173] U.S. Shrimp AB Report, *supra* note 151, para. 161.

[174] *Id.* (emphasis in original).

[175] *Id.* para. 165.

[176] *Id.* paras 166-172.

[177] U.S. Shrimp (Article 21.5) AB Report, *supra* note 144, para. 134.

[178] U.S. Shrimp AB Report, *supra* note 151, paras 173-174.

[179] *Id.* para. 177.

[180] *Id.* para. 144.

[181] *Id.* para. 180.

[182] *Id.* paras 178-83.

[183] Panel Report, *European Communities—Measures Affecting Asbestos and Asbestos-Containing Products*, para. 8.236, WT/DS135/R (September 18, 2000) (hereinafter, EC Asbestos Panel Report).

[184] *Id.*

[185] U.S. Gasoline AB Report, *supra* note 150, at 25.

[186] EC Asbestos Panel Report, *supra* note 179, paras. 8.237-8.239.

In: Carbon Leakage: Trade Issues in Emission Reduction ISBN: 978-1-61728-919-4
Editor: Victoria T. Greenwald

Chapter 3

CLIMATE CHANGE TRADE MEASURES: CONSIDERATIONS FOR U.S. POLICY MAKERS

Max Baucus

July 8, 2009

The Honorable Max Baucus
Committee on Finance, Chairman
United States Senate

Subject: *Climate Change Trade Measures: Considerations for U.S. Policy Makers*

Dear Mr. Chairman:

Changes in the earth's climate attributable to increased concentrations of greenhouse gases may have significant environmental and economic impacts in the United States and internationally. To mitigate climate change effects, countries are taking or considering varying approaches to reducing greenhouse gas emissions, such as carbon dioxide, which is the most important greenhouse gas due to its significant volume. These approaches range from measures to increase energy efficiency to longer-term efforts to develop technologies to establish a less carbon-intensive energy infrastructure. A U.S. policy to mitigate climate change may require domestic production facilities to achieve specified reductions or employ a market-based mechanism, such as establishing a price on emissions.

Between 2007 and 2009, Congress introduced a number of climate change bills, many of which contained proposals for a domestic emissions pricing system, such as a cap-and-trade system[1] or a carbon tax[2]. Through greenhouse gas emissions pricing, governments create an incentive for parties to lower their greenhouse gas emissions by placing a cost on them. However, imposing costs on energy-intensive industries in the United States could potentially place them at a disadvantage to foreign competitors. In addition, emissions pricing could have negative environmental consequences, such as "carbon leakage," whereby emissions reductions in the United States are replaced by increases in production and emissions in less-regulated countries. As the Congress considers the design of a domestic emissions pricing system, a key challenge will be balancing the need to reduce greenhouse gas emissions with the need to address the competitiveness of U.S. industries.

In anticipation of Senate deliberation on climate change legislative proposals, you asked us to examine how greenhouse gas emissions pricing could potentially affect the international competitiveness of U.S. industries, and to examine trade measures being considered as part of proposed U.S. climate change legislation.[3] On July 6, 2009, we briefed your staff covering: (1) what is known about estimating industry effects, (2) examples of industries that may be vulnerable to a loss in international competitiveness from emissions pricing, (3) trade measures and other approaches to address competitiveness issues, and (4) the potential international implications of trade measures. A copy of the slides presented at the briefing is attached.

To address these objectives, we interviewed officials and reviewed climate change literature and documents from relevant federal agencies, international organizations, policy institutes, businesses, professional organizations, and universities; reviewed and analyzed climate change legislation introduced between 2007 and 2009 and congressional hearing records; we reviewed and presented summary results for two studies that attempt to quantify the potential international competitiveness effects on domestic industries from greenhouse gas emissions pricing; and conducted interviews with officials from the embassies of Australia, Brazil, Canada, China, the European Union, and Mexico. We conducted our work from October 2008 to July 2009 in accordance with all sections of GAO's Quality Assurance Framework that are relevant to our objectives. The framework requires that we plan and perform the engagement to obtain sufficient and appropriate evidence to meet our stated objectives and to discuss any limitations in our work. We believe that the information and data obtained, and the analysis conducted, provide a reasonable basis for any findings and conclusions in this product. For additional details regarding our scope and methodology, see appendix I.

We are sending copies of this report to interested congressional committees and to other parties. In addition, the report will be available at no charge on GAO's Web site at http://www.gao.gov.

If you or your staff have any questions or wish to discuss this material further, please contact me at (202) 512-4347 or yagerl@gao.gov. Contact points for our Offices of Congressional Relations and Public Affairs may be found on the last page of this report. Christine Broderick (Assistant Director), Etana Finkler, Kendall Helm, Jeremy Latimer, Maria Mercado, and Ardith Spence, made significant contributions to this report. In addition, Karen Deans, David Dornisch, Grace Lui, and Jena Sinkfield provided technical assistance.

Sincerely yours,

Loren Yager
Director, International Affairs and Trade

WHY GAO DID THIS STUDY

Global climate change is one of the most significant long-term policy challenges facing the United States, and policies to mitigate climate change will have important economic, social, and environmental implications.

Members of Congress have introduced several bills to address the problem of climate change, many of which establish domestic emissions pricing by requiring firms that emit greenhouse gases either to pay a tax or to hold emission allowances. Whichever approach is taken, domestic emissions pricing could produce environmental benefits by encouraging U.S. firms to reduce their emissions of greenhouse gases. But such pricing could also harm U.S. firms' competitiveness, especially in energy-intensive industries where firms compete internationally. Additionally, there could be increased emissions abroad if production were to increase in other countries as a result of increased domestic costs of production resulting from a U.S. climate policy (carbon leakage). To help reduce impacts on U.S. firms and prevent carbon leakage, several climate change bills have also included trade measures or output-based rebates. The bills have included trade measures that would require importers to purchase emission allowances or pay a border tax for the greenhouse gas emissions associated with their imports. They have also designed output-based rebates to financially rebate industries for the costs incurred under a domestic emissions pricing system.

BRIEFING STRUCTURE

Background (slides 3-4)
Section 1: Estimating Industry Effects (slides 5-10)
Section 2: Potentially Vulnerable Industries (slides 11-20)
Section 3: Trade Measures (slides 21-31)
Section 4: International Trade Implications (slides 32-35)
Appendices (slides 36-46)

Estimating Industry Effects: Estimating the potential effects of domestic emissions pricing for industries in the United States is complex. If the United States were to regulate greenhouse gas emissions, production costs could rise for certain industries and could cause output, profits, or employment to fall. Within these industries, some of these adverse effects could arise through an increase in imports, a decrease in exports, or both. Estimates of adverse competitiveness effects are generally larger for industries that are both relatively energy and trade intensive. In 2007, these industries accounted for about 4.5 percent of domestic output.

Estimates of the effects vary because of key assumptions required by economic models. For example, models generally assume a price for U.S. carbon emissions, but do not assume a similar price by other nations. In addition, the models generally do not incorporate all policy provisions, such as legislative proposals related to trade measures and rebates that are based on levels of production.

Potentially Vulnerable Industries: Proposed legislation suggests that industries vulnerable to competitiveness effects should be considered differently. Industries for which competitiveness measures would apply are identified on the basis of their energy and trade intensity. Most of the industries that meet these criteria are in primary metals, nonmetallic minerals, paper, and chemicals, although significant variation exists for product groups (sub-industries) within each industry. Additional variation arises on the basis of the type of energy used and the extent to which foreign competitors' greenhouse gas emissions are regulated.

To illustrate variability in characteristics that make industries vulnerable to competitiveness effects, we selected example sub-industries within primary metals, non metallic minerals, paper products, and chemicals based on multiple factors. For example, we selected sub-industries that met both the energy and trade intensity criteria, examples that met only one criterion, and examples that met neither, but had significant imports from countries without greenhouse-gas pricing.

Trade Measures: Trade measures have been proposed to help address potential industry and environmental effects of a domestic emissions pricing system; however, questions exist about their proposed effectiveness. Supporters argue that trade measures may help prevent a decline in output by U.S. producers, prevent carbon leakage, and create leverage for other countries to reduce emissions. Opponents raise concerns that trade measures may motivate retaliatory actions, undermine efforts to secure multilateral consensus, and generate little leverage. In addition, potential implementation challenges may exist.

Objectives, Scope, and Methodology

GAO was asked to examine the potential effects of greenhouse gas emissions pricing on U.S. industries' international competitiveness and trade measures being considered as part of U.S. legislative proposals to address climate change.

Specifically, we examined: (1) what is known about estimating industry effects; (2) examples of industries that may be vulnerable to a loss in international competitiveness from emissions pricing; (3) trade measures and other approaches to address competitiveness issues; and (4) potential international implications of trade measures.

To address these objectives, we interviewed officials and reviewed climate change literature and documents from U.S. agencies, international organizations, policy institutes, and professional organizations; reviewed and analyzed climate change legislation introduced between 2007 and 2009 and congressional hearing records; reviewed and presented summary results for two studies attempting to quantify the potential international competitiveness effects on domestic industries from emissions pricing; and conducted interviews with officials from the embassies of Australia, Brazil, Canada, China, the European Union, and Mexico.

The Office of the United States Trade Representative provided technical comments on this report.

The analysis in this report reflects changes to the American Clean Energy and Security Act of 2009 (HR 2454), as of June 26, 2009.

For additional details regarding our scope and methodology, see appendix I.

Output-based rebates are measures that offset incurred costs to industries due to emissions pricing and can take the form of a per-unit rebate, allocation of allowances or tax credit. These rebates have been proposed to address competitiveness effects; however, limitations with these measures may also exist. For example, output-based rebates may increase costs for industries that do not receive them, under certain conditions.

How a trade measure is ultimately designed will have important implications for the effectiveness of the overall measure in addressing industry and environmental effects. For example, delaying the timing for when the trade measure goes into effect could result in the measure being a more effective leveraging tool, and would provide other countries an opportunity to implement similar carbon mitigation policies to reduce global emissions. On the other hand, having the trade measure go into effect simultaneously when the domestic cap-and-trade system goes into effect may reduce impacts on U.S. industry competitiveness

International Trade Implications: A unilateral U.S. trade measure could have important international implications on U.S. bilateral and multilateral trade relations. For example, other countries could view U.S. trade measures as trade restrictions or sanctions, which could lead them to implement restrictions against U.S. exports. Attention also has focused on the potential for trade measures or output-based rebates to be challenged under World Trade Organization (WTO) rules.

Assessing the international trade implications is difficult for a number of reasons. One is that it depends in part upon how other nations reduce their carbon emissions, and whether they perceive any U.S. measures as likely to affect their exports. In addition, the outcome of any WTO challenge, if any, would be uncertain and may depend on how the measure is implemented.

The Greenhouse Effect

Greenhouse gases, such as carbon dioxide, methane, nitrous oxide, and other gases, increase temperatures by trapping heat that would otherwise escape the earth's atmosphere. The heat-trapping effect, known as the greenhouse effect, moderates atmospheric temperatures, keeping the earth warm enough to support life. Although each unit of non-carbondioxide greenhouse gas generally has a greater warming effect than each unit of carbon dioxide, carbon dioxide is the most important greenhouse gas because of its significant volume. The potency of other greenhouse gases may be expressed in terms of their warming potential relative to carbon dioxide, or their carbon dioxide equivalent.

Estimated Growth in Emissions

According to the IPCC, in 2004, developed countries, including the United States, constituted 20 percent of the global population, but were responsible for nearly half of global greenhouse gas emissions. However, the IPCC projects that between 2000 and 2030, two-thirds to three-quarters of the projected increase in global carbon dioxide emissions will occur in developing countries. The IPCC also projects that compared with 2000, global greenhouse gas emissions will increase between 25 percent and 90 percent by 2030 in the absence of policies to reduce greenhouse gas emissions worldwide.

BACKGROUND

Greenhouse Gas Emissions

The Intergovernmental Panel on Climate Change (IPCC) was established to provide scientific and objective information on climate change. According to the IPCC, global atmospheric concentrations of greenhouse gases have increased markedly as a result of human activities over the past 200 years, contributing to a warming of the earth's climate. Climate change is a long-term and global issue because greenhouse gases disperse widely in the atmosphere once emitted and can remain there for an extended period of time. Among other potential impacts, climate change could threaten coastal areas with rising sea levels, alter agricultural productivity, and increase the intensity and frequency of floods and tropical storms. While the effect of increases in greenhouse gas emissions on ecosystems and economic growth is expected to vary across regions, countries, and economic sectors, a panel of 18 climate change economists convened by GAO in coordination with the National Academy of Sciences agreed that the Congress should consider using a market-based mechanism to place a price on emissions, and 14 of the 18 panelists were at least moderately confident that the benefits of taking these actions would outweigh the costs. See GAO-08-605.

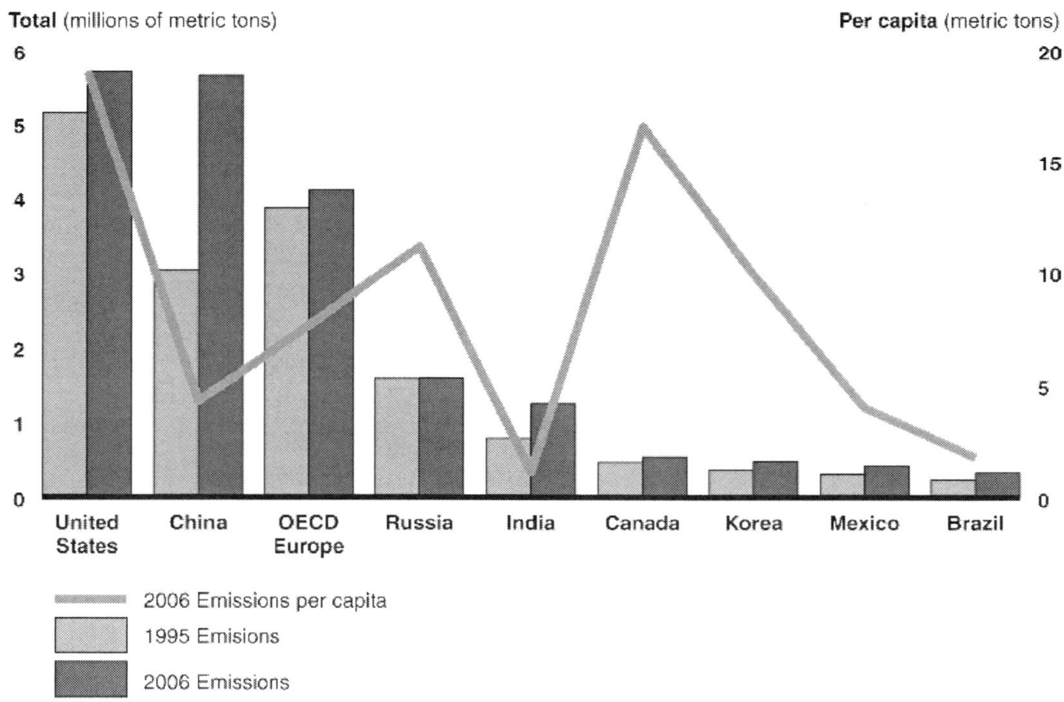

Source: International Energy Agency data on carbon dioxide emissions from fuel consumption.

Figure 1. Data on Global Carbon Dioxide Emissions

As shown in figure 1, the United States emitted more carbon dioxide than any other country in 2006, while China recently surpassed the European countries that are members of the Organization for Economic Cooperation and Development (OECD) as the world's second largest emitter. On a per capita basis, the United States and Canada rank first and second in carbon dioxide emissions. Between 1990 and 2006, carbon dioxide emissions grew substantially in both China and India.

The Kyoto Protocol

The Kyoto Protocol, adopted in Kyoto, Japan, and ratified by more than 180 countries, sets legally binding emissions targets for 37 industrialized countries and the European Community (Annex I countries).

The Bali Action Plan

The year 2012 will mark the end of the first commitment period of the Kyoto Protocol. In Bali, Indonesia, in 2007, countries agreed to the Bali Action Plan, which defined a process and timeframe to enable the full effective and sustained implementation of the Convention through long-term cooperative action up to and beyond 2012. According to the Bali Action Plan, the negotiations process is scheduled to conclude in December 2009 at a major summit in Copenhagen, Denmark.

Existing Emissions Pricing Systems

Greenhouse gas emissions pricing has been implemented in several other cases, including a cap-andtrade system in the European Union (EU) and carbon tax systems in Sweden, Finland, Norway, and the Netherlands. The world's largest emissions pricing system, the European Union Emissions Trading Scheme (EU ETS), covers CO_2 emissions from more than 11,000 energy-intensive installations. The EU ETS began with a pilot phase that ran from 2005 to 2007. Its second trading period is currently under way. For more information see GAO-09-151.

International Negotiations and Emissions Pricing

Greenhouse gas emissions reductions in all major countries will be required to stabilize greenhouse gas concentrations at a level that would prevent dangerous climate change. In 1992, 192 countries, including the United States, joined the United Nations Framework Convention on Climate Change (UNFCCC), an international treaty to consider what may be done to reduce and adapt to global warming. In 1998, the United States signed the Kyoto Protocol, an international agreement to specifically limit greenhouse gas emissions. The United States is not bound by the protocol's terms because it was not ratified by the Senate. Both the UNFCCC and the Kyoto protocol operate under the principal of "common but differentiated responsibilities" to reflect the agreement that each nation must contribute to addressing climate change, but that the magnitude of its efforts should differ according to national circumstances. For example, under the protocol, only Annex I countries have quantified emission reduction obligations. To negotiate the next round of international commitments, parties to the UNFCCC agreed to the "Bali Action Plan" in December 2007.

Countries can take varying approaches to reducing greenhouse gas emissions. Since energy use is a significant source of greenhouse gas emissions, policies designed to increase energy efficiency or induce a switch to less greenhouse gas-intensive fuels, such as from coal to natural gas, can reduce emissions in the short term. In the long term, however, major technology changes will be needed to establish a less carbon-intensive energy infrastructure. To that end, a U.S. policy to mitigate climate change may require facilities to achieve specified reductions or employ a market-based mechanism, such as establishing a price on emissions.

Proposed Legislation Includes Two Options for Domestic Emissions Pricing Systems

Through greenhouse gas emissions pricing, governments create an incentive for parties to lower their greenhouse gas emissions by placing a cost on the emissions. Proposed U.S. legislation includes two basic approaches to greenhouse gas emissions pricing: a carbon tax system and a cap-and-trade system. A carbon tax system would impose a fee on emissions of greenhouse gases. Under a cap-and-trade system, the government would cap greenhouse gas emissions, limiting the total quantity of emissions from regulated sources under the system. The government would also issue allowances, or permits to emit greenhouse gases, equal to the overall cap or quota. Regulated sources would be required to hold enough allowances to cover their emissions, and allowances could be bought and sold as needed to meet the system's requirements in the least expensive manner.

Several bills to implement emissions pricing in the United States have been introduced in the 110th and the 111th Congresses. These bills have included both cap-and-trade and carbon tax proposals. Some of the proposed legislation also include measures intended to limit potentially adverse impacts on the international competitiveness of domestic firms.

1. Estimating Industry Effects

Introduction

In this section, we present information on what is known about how U.S. emissions pricing could potentially affect industry competitiveness. Importantly, estimating the effects of domestic emissions pricing for industries in the United States is complex. For example, if the United States were to regulate greenhouse gas emissions, production costs could rise for many industries and could cause output, profits, or employment to fall. However, the magnitude of these potential effects is likely to depend on the greenhouse gas intensity of industry output and on the domestic emissions price, which is not yet known, among other factors. Additionally, if U.S. climate policy were more stringent than in other countries, some domestic industries could experience a loss in international competitiveness. Within these industries, adverse competitiveness effects could arise through an increase in imports, a decrease in exports, or both.

To estimate the potential economic effects of U.S. emissions pricing on domestic industries, models require key assumptions, which cause estimates to vary. In this section, we review two economic models that specifically estimate competitiveness effects for U.S. industries. In each, larger adverse effects occur for those U.S. industries that are both relatively energy and trade intensive. We also present information on the potential for adverse international competitiveness effects to be a source of carbon leakage.

U.S. Emissions Pricing Effects on Production Costs and Output

For regulated sources, greenhouse gas emissions pricing would increase the cost of releasing greenhouse gases. As a result, it would encourage some of these sources to reduce their emissions, compared with business-as-usual. Under domestic emissions pricing, production costs for regulated sources could rise as they either take action to reduce their emissions or pay for the greenhouse gases they release. Cost increases are likely to be larger for production that is relatively greenhouse gas intensive, where greenhouse gas intensity refers to emissions per unit of output. Cost increases may reduce industry profits, or they may be passed on to consumers in the form of higher prices. To the extent that cost increases are passed on to consumers, they could demand fewer goods, and industry output could fall.

While emissions pricing would likely cause production costs to rise for certain industries, the extent of this rise and the resulting impact on industry output are less certain due to a number of factors. For example, the U.S. emissions price and the emissions price in other countries are key variables that will help to determine the impact of emissions pricing on domestic industries. However, future emission prices are currently unknown. Furthermore, to the extent that emissions pricing encourages technological change that reduces greenhouse gas intensity, potential adverse effects of emissions pricing on profits or output could be mitigated for U.S. industries. Additionally, policy options such as output-based rebates could also offset some of the added costs of emissions pricing for some firms but involve tradeoffs, such as possibly increasing costs for firms that do not receive them.

Estimated Effects from Economic Models of Emissions Pricing

Several studies by U.S. agencies and experts have used models of the economy to simulate the effects of emissions pricing policy on output and related economic outcomes. These models generally find that emissions pricing will cause output, profits or employment to decline in sectors that are described as energy intensive, compared with business-as-usual. In general, these studies conclude that these declines are likely to be greater for these industries, as compared to other sectors in the economy. However, some research suggests that not every industry is likely to suffer adverse effects from emissions pricing. For example, a long-run model estimated by Ho, Morgenstern, and Shih (2008) predicts that some U.S. sectors, such as services, may experience growth in the long run as a result of domestic emissions pricing. This growth would likely be due to changes in consumption patterns in favor of goods and services that are relatively less greenhouse gas intensive.

Potential international competitiveness effects depend in part on the stringency of U.S. climate policy relative to other countries. For example, if domestic greenhouse gas emissions pricing were to make emissions more expensive in the United States than in other countries, production costs for domestic industries would likely increase relative to their international competitors, potentially disadvantaging industries in the United States. As a result, some domestic production could shift abroad, through changes in consumption or investment patterns, to countries where greenhouse gas emissions are less stringently regulated. For example, consumers may substitute some goods made in other countries for some goods made domestically. Similarly, investment patterns could shift more strongly in favor of new capacity in countries where greenhouse gas emissions are regulated less stringently than in the United States.

Economists such as Aldy and Pizer (2009) have defined adverse international competitiveness effects as the part of a decline in domestic output that is due to a shift of production abroad in response to emissions pricing, or the part of a decline in domestic output that not also matched by a decline in domestic consumption. For example, if domestic output were to fall by more than domestic consumption, the difference could be explained by changes in trade patterns through a reduction in exports, an increase in imports, or both.

Factors that May Affect International Competitiveness

Stakeholders and experts have identified two criteria, among others, that are important in determining potential vulnerability to adverse competitiveness effects: trade intensity and energy intensity. Trade intensity is important because international competitiveness effects arise from changes in trade patterns. For example, if climate policy in the United States were more stringent than in other countries, international competition could limit the ability of domestic firms to pass increases in costs through to consumers. Energy intensity is important because the combustion of fossil fuels for energy is a significant source of greenhouse gas emissions, which may increase production costs under emissions pricing. International competitiveness effects may also depend on other factors, including transportation costs and access to markets for natural resources, capital, or labor.

Industries that are both relatively trade intensive and energy intensive include primary metals, nonmetallic minerals, paper products, and chemicals. These four industries provided around 4.5 percent of total U.S. output in 2007.

Legislation passed in June 2009 by the House of Representatives, H.R. 2454, 111th Cong. (2009), uses the criteria of trade intensity and energy intensity or greenhouse gas intensity, among others, to determine eligibility for the Emission Allowance Rebate Program that is part of the legislation. H.R. 2454 specifies how to calculate the two criteria. Trade intensity is defined as the ratio of the sum of the value of imports and exports within an industry to the sum of the value of shipments (output) and imports within the industry. Energy intensity is defined as the industry's cost of purchased electricity and fuel costs, or energy expenditures, divided by the value of shipments (output) of the industry.

The energy intensity measure specified in this legislation, however, might not be an ideal measure of vulnerability. First, the relationship between energy expenditures and emissions from energy use is indeterminate. For example, energy expenditures depend on the price of fuel and on the quantity of fuel used, while emissions from energy use depend on the quantity and type of fuel used. Additionally, an increase in energy prices may cause expenditures on energy to rise and energy use to fall, which could also cause emissions from energy use to fall. Second, the energy intensity measure might not reflect the extent to which substitutes for energy are available. For example, regulated sources may be able to reduce their energy intensity or switch to energy sources that are less greenhouse gas intensive. To the extent that less greenhouse gas intensive substitutes are available, some industries could reduce the effect of emissions pricing on their costs and on their competitiveness.

Key Assumptions of Models That Estimate Competitiveness Effects

Key assumptions made by economic models of emissions pricing include, among others, the domestic emissions price, the emissions price in other countries, the level of industry aggregation, and the time horizon during which producers may adjust their production methods or energy sources in response to emissions pricing.

- *Domestic emissions price:* The domestic emissions price could affect the size of a potential increase in production costs from emissions pricing. In particular, increases in production costs are expected to be larger when the domestic emissions price is greater. Under a carbon tax system, the domestic emissions price would equal the tax per unit of emissions. Under a cap-and-trade system, the domestic emissions price would equal the market-clearing allowance price, at which the quantity of allowances supplied equals the quantity of allowances demanded.
  The domestic emissions price could also depend on cost containment features of an emissions pricing system, such as banking, borrowing or carbon offsets, among other features.
- *Emissions price in other countries*: Together with the domestic emissions price, the emissions price in other countries will help to determine the relative stringency of climate policy in the United States. When the difference between the domestic emissions price and a lower emissions price in other countries is larger, U.S. industries may be more likely to experience adverse competitiveness effects from domestic emissions pricing. Under a cap-and-trade system, differences in the allowance price between two countries may dissipate if international trading of allowances is allowed.
- *Level of aggregation*: The range of estimated effects may depend on the level of industrial or sectoral aggregation used. For example, estimates for highly aggregated sectors could mask more extreme variation at the more narrowly aggregated sub-industry level.
- *Time horizon*: In models of economic activity, the amount of time considered after policy implementation may affect model results. For example, the time horizon may affect the ability of firms to pass through added costs to consumers in the form of higher prices, change their production processes, or develop and adopt new technologies. A longer time horizon could also allow for investments in new capacity.
- *Policy Scenarios*: Model results may depend in part upon the range of policy scenarios incorporated. For example the incorporation of trade measures and output-based rebates may affect the results.

Estimates of competitiveness effects are uncertain

Table 1 presents results from two studies that attempt to quantify the potential competitiveness effects on domestic industries from emissions pricing. For each study, a range of estimates is reported for two effects: (1) the potential decline in domestic, energy-intensive manufacturing output that is due to U.S. emissions pricing and (2) the part of this decline that is due to adverse international competitiveness effects. The adverse competitiveness effects are computed as the difference between the production and consumption impacts of emissions pricing. For example, if production declines by 3.7 percent and consumption declines by 2.9 percent as a share of production, the adverse competitiveness effect equals the -0.8 percent change in production that is not matched by a change in domestic consumption.

Table 1. Estimates of Emissions Pricing Effects on Energy-Intensive Manufacturing Industries in the United States

	Study	
	Aldy and Pizer (2009)	**Ho, Morgenstern, and Shih (2008)**
Estimated decline in industry output due to emissions pricing (range of estimates)	1.0% – 4.3%	0.91% - 1.30%
Part of estimated decline in industry output due to adverse international competitiveness effects (range of estimates)	0.3% – 1.8%	0.42% - 0.68%
Domestic emissions price (per ton of carbon dioxide)	$15	$10
Emissions price in other countries (per ton of carbon dioxide)	$0	$0
Time horizon	1 year to a few years	Long run
Number of industries in model	400+	21
Number of industries for which results are reported by authors	55 energy-intensive	21
Number of industries included in the ranges of estimates reported above	55	3 most energy-intensive
Incorporation of trade measure and output-based rebate policies	no	no

Source: GAO analysis of results reported by Aldy and Pizer (2009) and Ho, Morgenstern, and Shih (2008). Note: The number of industries in both reflects, in part, the level of industry aggregation. While Aldy and Pizer (2009) examine more than 400 industries at a more narrowly aggregated level, they report estimates only for 55 energy-intensive industries among the more than 400 industries they examine. Ho et al. (2008) examine industries at a more aggregated level and report estimates for all 21 manufacturing and nonmanufacturing sectors that they consider. Ho et al. (2008) also report estimates of the effects of emissions pricing on industry output for multiple time horizons, but they report both production and consumption effects—both of which are necessary to compute competitiveness effects—only for their long run estimates. For the study by Aldy and Pizer (2009), the estimated decline in industry output and the part that is due to competitiveness effects are reported as a percentage of industry output. For the study by Ho et al. (2008), the estimated decline in industry output and the part that is due to competitiveness effects are reported as a share of sectoral consumption.

In both studies, estimates of competitiveness effects are generally greater for industries that are relatively trade and energy intensive.

The estimated impacts depend on key assumptions of the models used to generate these results. Both studies assume unilateral action by the United States, or an emissions price of $0 per ton of carbon dioxide in the rest of the world, which may overstate the estimates of adverse competitiveness effects for U.S. industries to the extent that other countries also regulate greenhouse gas emissions.

Aldy and Pizer's (2009) model allows for some adjustments and is consistent with a time horizon of 1 year to a few years. Ho et al. (2008) generate their long-run estimates using a model that allows for more adjustments.

International Competitiveness Effects and Carbon Leakage

Reducing carbon emissions in the United States could result in carbon leakage through two potential mechanisms. First, if domestic production were to shift abroad to countries where greenhouse gas emissions are not regulated, emissions in these countries could grow faster than expected otherwise. Through this mechanism, some of the expected benefits of reducing emissions domestically could be offset by faster growth in emissions elsewhere, according to Aldy and Pizer (2009).

Carbon leakage may also arise from changes in world prices that are brought about by domestic emissions pricing. For example, U.S. emissions pricing could cause domestic demand for oil to fall. Because the United States is a relatively large consumer of oil worldwide, the world price of oil could fall when the U.S. demand for oil drops. The quantity of oil consumed by other countries would rise in response, increasing greenhouse gas emissions from the rest of the world. These price effects may be a more important source of carbon leakage than the trade effects described above. See Fischer and Fox (2009) and EPA (2009).

2. Potentially Vulnerable Industries

Introduction

Energy and trade intensities of the four manufacturing industries, in the aggregate, generally meet both vulnerability criteria in proposed legislation. As shown along the right axis in figure 2, nonmetallic minerals is the most energy intensive at 6.1 percent in 2006. As shown along the left axis, chemicals is the most trade intensive, at 45 percent in 2007. While chemicals does not have an energy intensity of 5 percent or greater – and is located outside of the shaded area— several subindustries within chemicals meet that vulnerability criterion.

Together, these four industries provided 23 percent of total U.S. manufacturing output in 2007 and had trade flows of about $500 billion. As shown by the size of the column, chemicals is the largest industry, with output of $664 billion in 2007. Accordingly, chemicals also accounted for the largest share of carbon dioxide emissions from manufacturing, at 22 percent in 2002. However, the estimated greenhouse gas intensity of chemical products— the level of carbon dioxide emitted per unit of energy—was less than that for both primary metals and nonmetallic minerals in 2002.

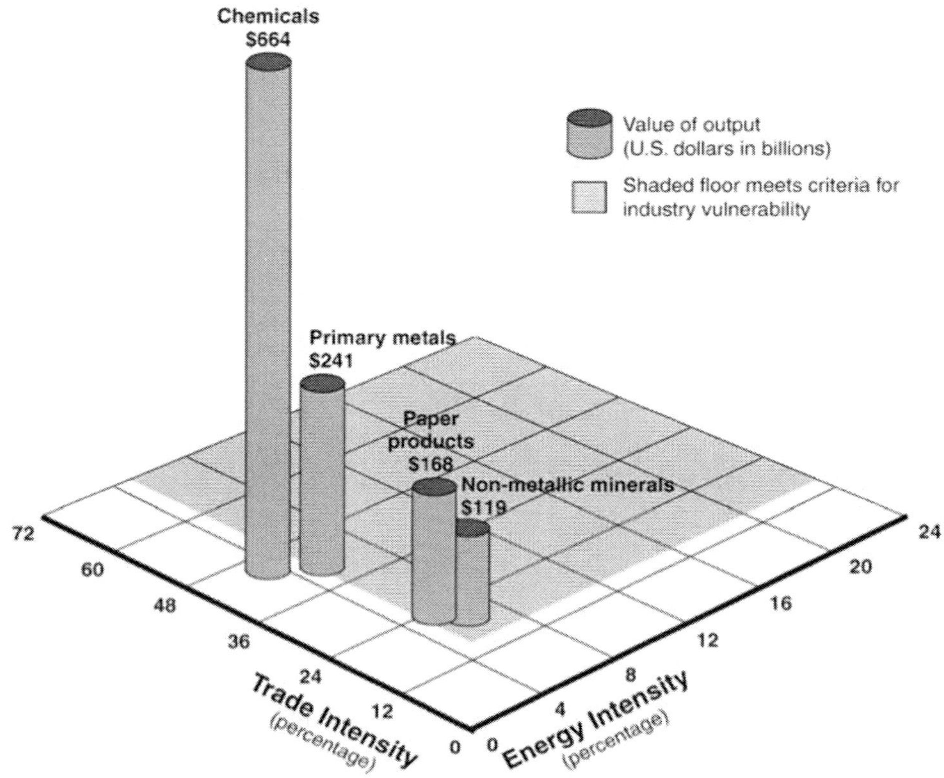

Source: GAO analysis of Department of Energy emissions data and Department of Commerce energy data for 2006 and trade and output data for 2007.

Notes to figure 2

Figure 2. Energy and Trade Intensity Are Indicators of Vulnerability

In this section, we provide some information on certain U.S. industries that, by paying a price on greenhouse gas emissions, could potentially lose competitiveness compared with foreign industries without comparable climate policies. Among many factors, two key indicators of potential vulnerability to adverse competitiveness effects are an industry's energy intensity and trade intensity. Proposed U.S. legislation specifies that: (a) either an energy intensity or greenhouse gas intensity of 5 percent or greater; and (b) a trade intensity of 15 percent or greater be used as criteria to identify industries for which trade measures or rebates would apply. Since data on greenhouse gas intensity is less complete, we focus our analysis on industry energy intensity. Most of the industries that meet these criteria are in primary metals, nonmetallic minerals, paper, and chemicals, yet there is significant variation in specified vulnerability characteristics among different product groups ("sub-industries"). The following pages include examples of this variation, as well as information on the type of energy used and location of import and export markets. Data shown are for the latest year available, with additional discussion of proposed vulnerability criteria in appendix I and industry information in appendix II.

2.1. Primary Metals Examples

As shown by sub-industry examples in figure 3, energy and trade intensities differ within primary metals. For example, primary aluminum meets the vulnerability criteria with an energy intensity of 24 percent and trade intensity of 62 percent. Ferrous metal foundries meets the energy intensity criteria, but not the trade intensity criteria. Steel manufacturing—products made from purchased steel—and aluminum products fall short of both vulnerability criteria. Iron and steel mills has an energy intensity of 7 percent and a trade intensity of 35 percent and is by far the largest sub- industry example, with a 2007 value of output of over $93 billion.

Among the primary metals sub- industry examples, types of energy used also vary. Iron and steel mills use the greatest share of coal and coke, and steel manufacturing and ferrous metal foundries use the greatest proportion of natural gas. Since coal is more carbon intensive than natural gas, sub-industries that rely more heavily on coal could also be more vulnerable to competitiveness effects. The carbon intensity of electricity, used heavily in the production of aluminum, will also vary on the basis of source of energy used to generate it and the market conditions where it is sold. Data shown for "aluminum" include primary aluminum and aluminum products and net electricity is the sum of net transfers plus purchases and generation minus quantities sold.

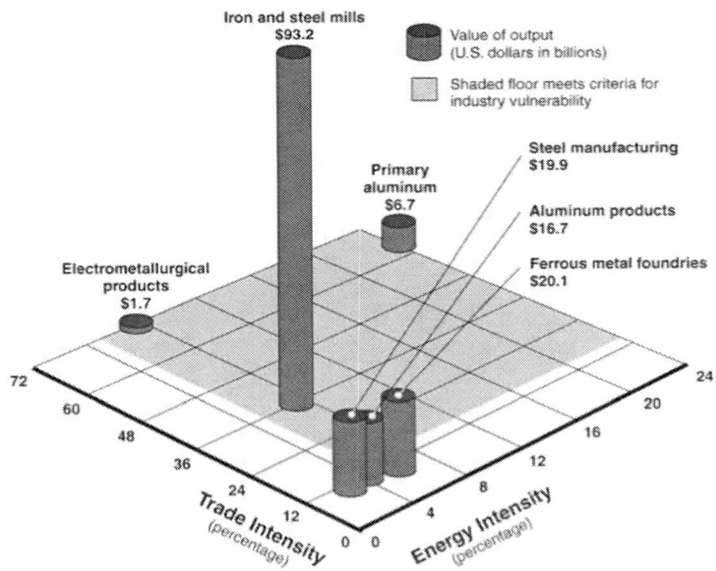

Source: GAO analysis of Department of Commerce energy data for 2006 and trade data for 2007.

Notes to figure 3

Figure 3. Energy and Trade Intensity Indicators Vary by Sub- Industry

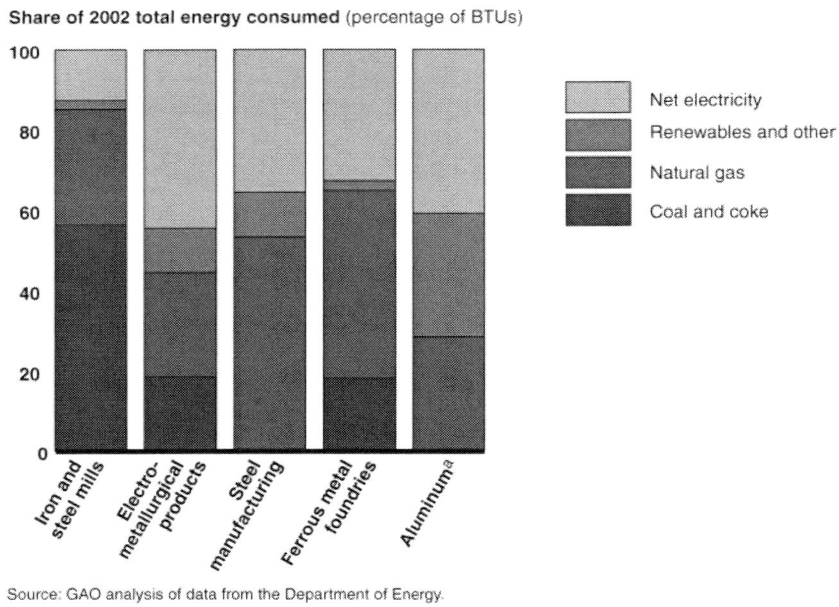

Notes to figure 4

Figure 4. Type of Energy Used Is Important for Carbon Intensity

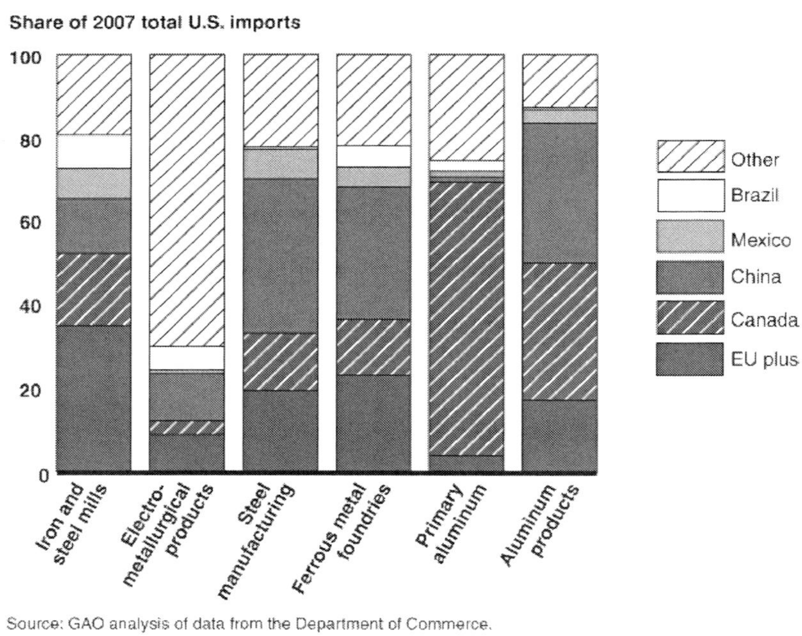

Notes to figure 5

Figure 5. Source of Imports Is Important for Trade Intensity

Industry vulnerability may further vary depending on the share of trade with countries that do not have carbon pricing. To illustrate this variability, figure 5 provides data on the share of imports by source, since imports exceed exports in each of the primary metals examples. As shown, while primary aluminum is among the most trade intensive, the majority of imports are from Canada, an Annex I country with agreed emission reduction targets. For iron and steel mills, over one- third of imports are from the EU and other Annex I countries, not including Canada ("EU plus"). However, for iron and steel mills, almost 30 percent of imports are also from the non-Annex I countries of China, Mexico, and Brazil. While less trade intensive, steel manufacturing and aluminum products each has greater than one- third of imports from China alone

Adverse competitiveness effects from emissions pricing could increase the already growing share of Chinese imports that exists in some of the sub-industries. Among the examples, iron and steel mills, steel manufacturing, and aluminum products exhibit a growing trade reliance on Chinese imports since 2002. This trend has largely been driven by lower labor and capital costs in China and, according to representatives from the steel industry, China has recently been producing 50 percent of the world's steel.

2.2. Nonmetallic Minerals Examples

Among nonmetallic minerals sub- industry examples, cement has the highest energy intensity, at 16 percent, and glass and clay building materials have the highest trade intensity, at 36 percent and 35 percent, respectively. While, concrete is the largest sub-industry, with a value of output of over $30 billion in 2007, it fails to meet both of the vulnerability criteria and thus lies outside of the shaded floor area. Lime and gypsum meets the energy intensity criteria but fails to meet the trade intensity criteria.

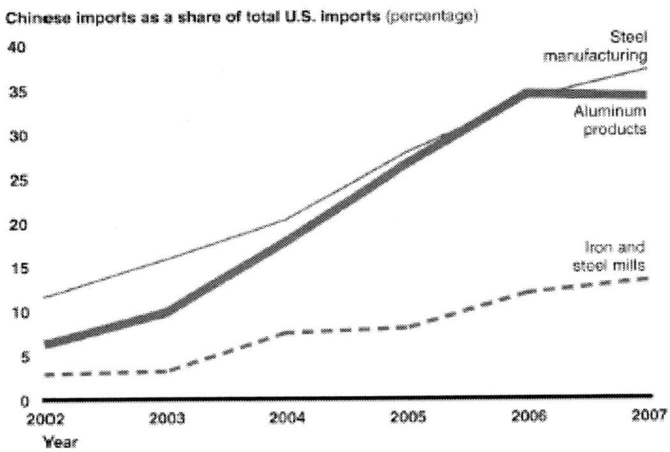

Total U.S. imports by value (U.S. dollars in millions)						
Year	2002	2003	2004	2005	2006	2007
Iron and steel mills	$12,558	10,808	23,355	25,131	33,060	30,445
Steel manufacturing	$1,110	1,184	1,792	1,884	1,916	1,822
Aluminum products	$498	549	718	913	1,209	1,183

Source: GAO analysis of data from the Department of Commerce.

Notes to figure 6

Figure 6. Sub-Industries With Growing Share of Imports from China

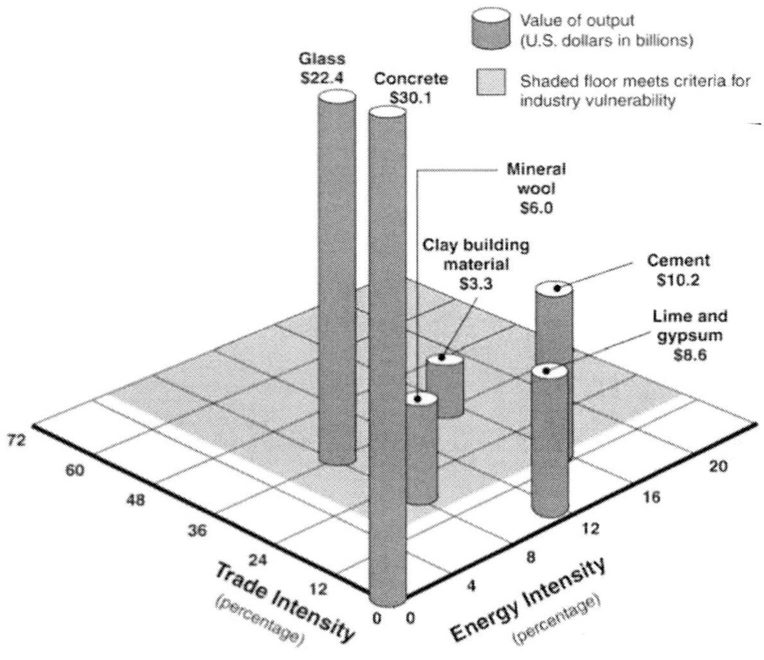

Notes to figure 7

Figure 7. Energy and Trade Intensity Indicators Vary by Sub- Industry

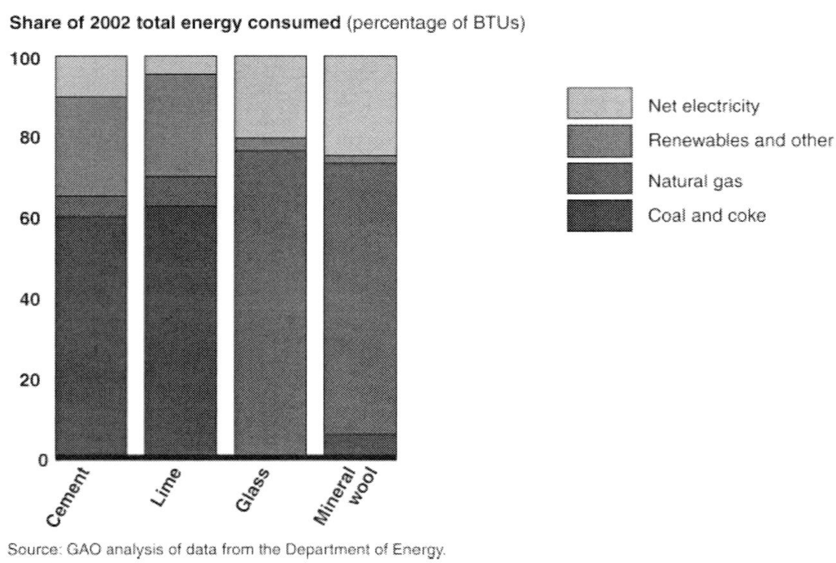

Notes to figure 8

Figure 8. Type of Energy Used Is Important for Carbon Intensity

Cement and lime rely on coal and coke sources for about 60 percent of energy consumed, while glass and mineral wool rely on natural gas for over 65 percent of energy consumed. Since coal is more carbon intensive than natural gas, sub-industries that rely more heavily on coal could also be more vulnerable to competitiveness effects. Data on energy use by the sub-industry examples excluded from the chart are not currently reported by the Department of Energy.

For the nonmetallic mineral examples, total imports exceed total exports. As shown in figure 9, glass has the largest share of imports from China, at 26 percent. Another 17 percent of glass imports are sourced from Mexico. For the other trade- intensive sub-industries of cement, clay building material, and mineral wool, about 30 percent of imports are from China, Mexico, and Brazil. Conversely, nearly 60 percent of lime and gypsum imports are from Canada.

The one sub-industry where exports comprise more than 60 percent of trade flows is concrete, with 30 percent of concrete exports sold to markets in Mexico and the Middle East.

Since 2002, a trend of increasing Chinese imports as a share of total imports is evident for glass and mineral wool. While the share of cement imports from China declined from 2006 to 2007, reliance on Chinese imports has also grown relative to 2002. According to representatives of the U.S. cement industry, Chinese production costs are between 20 percent and 40 percent of those in the United States.

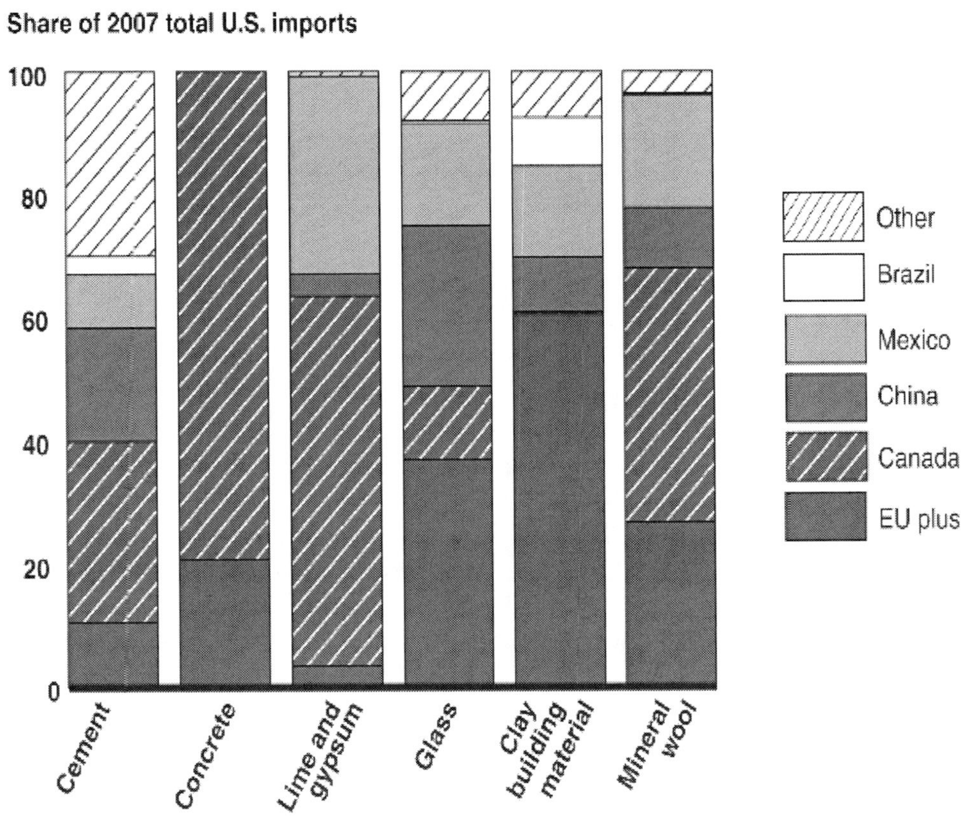

Source: GAO analysis of data from the Department of Commerce.

Notes to figure 9

Figure 9. Source of Imports Is Important for Trade Intensity

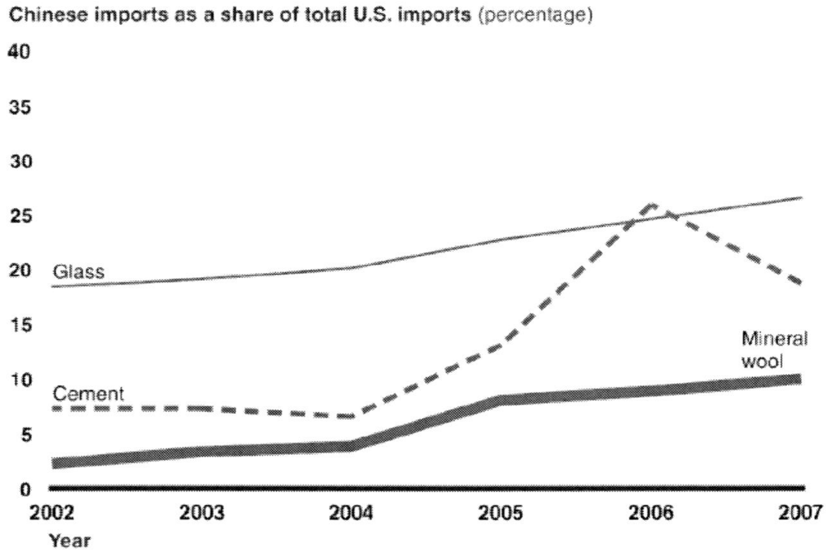

Total U.S. imports by value (U.S. dollars in millions)						
Year	2002	2003	2004	2005	2006	2007
Cement	$938	924	1,140	1,567	1,837	1,325
Glass	$4,293	4,436	4,947	5,125	5,491	5,726
Mineral wool	$332	360	448	510	577	489

Source: GAO analysis of data from the Department of Commerce.

Notes to figure 10

Figure 10. Sub-Industries With Growing Share of Imports from China

2.3. Paper Products Examples

Sub-industry examples within paper products also show variation in energy and trade intensity. Pulp mills is the most trade intensive, at 98 percent. Paper mills is the largest sub-industry, with a 2007 value of output of over $51 billion and its trade intensity is 30 percent. Pulp mills and paper mills each has an energy intensity of 8 percent and, thus, meet both vulnerability criteria. While paperboard mills has an energy intensity of 12 percent, it does not meet the trade intensity criteria. Paperboard containers fails to meet both criteria but was the second largest sub-industry in 2007, with a value of output of over $47 billion.

In paper mills and paperboard mills, the largest share of energy consumed is from the relatively less carbon-intensive renewable and other (such as fuel oil and steam) energy sources. An additional 20 percent of energy consumed is from natural gas, while coal and coke sources provide 14 percent of energy for paper mills and 9 percent of energy for paperboard mills. While energy use data for pulp mills is reported by the Department of Energy, consumption of coal and coke, specifically, is not publicly available, so percentage shares could not be computed. Data on energy use by paperboard containers are not currently reported.

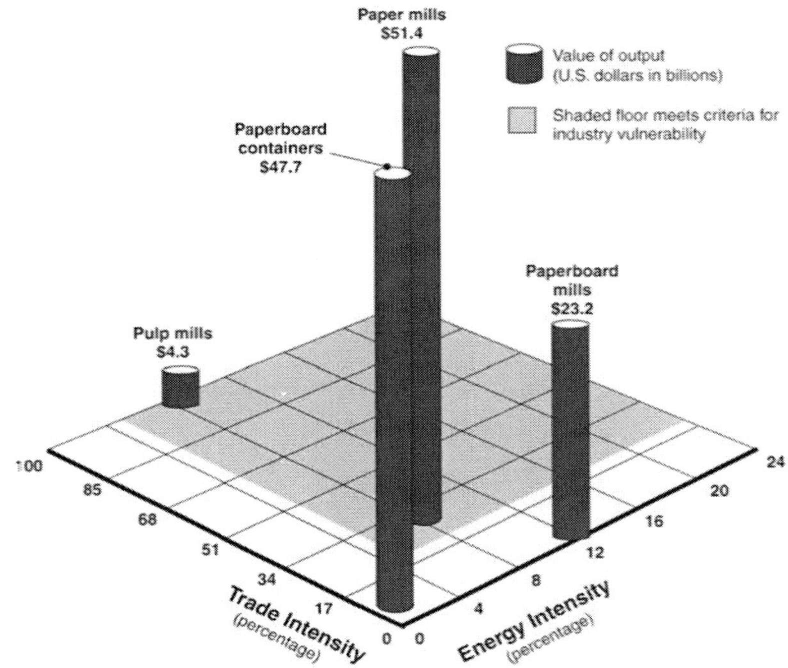

Source: GAO analysis of Department of Commerce energy data for 2006, output data for 2006 and 2007, and trade data for 2007.

Notes to figure 11

Figure 11. Energy and Trade Intensity Indicators Vary by Sub- Industry

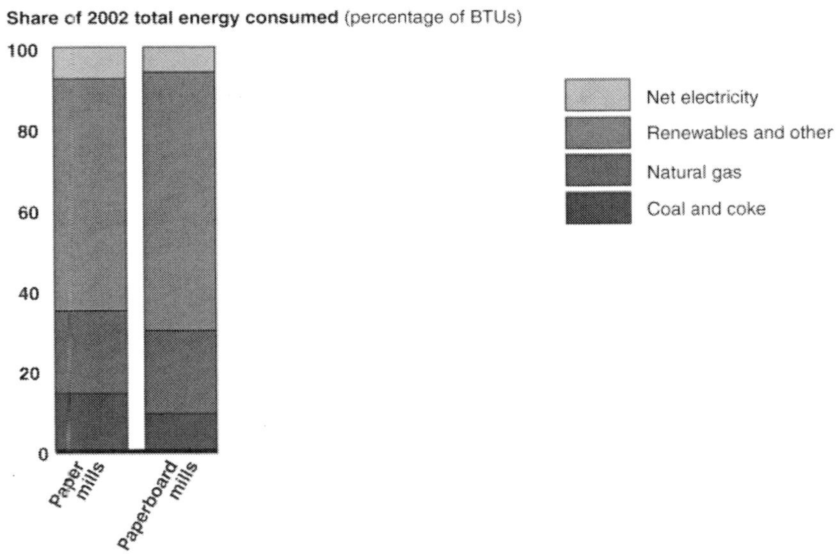

Source: GAO analysis of data from the Department of Energy.

Notes to figure 12

Figure 12. Type of Energy Used Is Important for Carbon Intensity

For the paper products sub-industry examples, total imports exceed total exports. For each, the largest share of imports is from Canada, an Annex I country with agreed emission reduction targets. However, for the most trade-intensive example of pulp mills, 19 percent of imports are from Brazil. While paperboard containers has a trade intensity of only 6 percent, 27 percent of imports are from China.

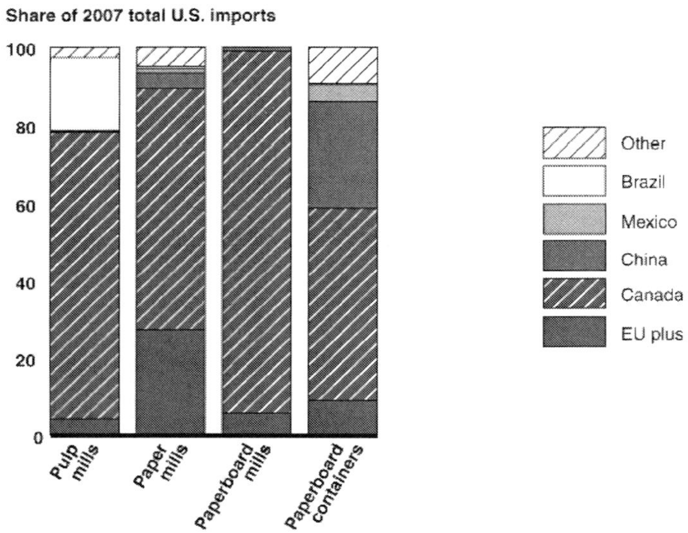

Source: GAO analysis of data from the Department of Commerce.

Figure 13. Source of Imports Is Important for Trade Intensity

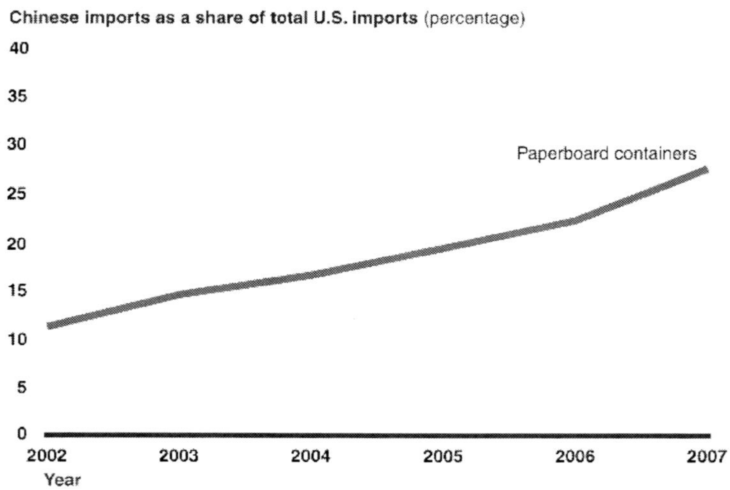

Total U.S. imports by value (U.S. dollars in millions)						
Year	2002	2003	2004	2005	2006	2007
Paperboard container	$790	875	926	1,033	1,187	1,275

Source: GAO analysis of data from the Department of Commerce.

Notes to figure 14

Figure 14. Sub-Industries With Growing Share of Imports from China

Among the paper products sub-industries, paperboard containers is the only example with a significant share of imports from China. While paperboard containers fails to meet the trade-intensity vulnerability criteria, emissions pricing could magnify the growing share of imports from China for that sub- industry. From 2002 to 2007, this share has grown from 11 percent to 27 percent of U.S. imports.

2.4. Chemicals Examples

Among chemicals examples, alkalies and chlorine is the most energy intensive, at 26 percent, and nitrogenous fertilizers is the most trade intensive, at 82 percent. Except for industrial gases, each of the examples meets the energy and trade intensity criteria for vulnerability. Industrial gases, the largest sub-industry example, with a 2007 value of output of almost $9 billion, has an energy intensity of 15 percent but a trade intensity of only 5 percent. Not shown in figure 15 is petrochemicals, with a 2007 value of output of almost $62 billion. The Department of Commerce does not publicly provide energy-intensity data for petrochemicals, though industry experts estimate it to be at above 5 percent but below 20 percent. The year 2007 trade data indicate a petrochemicals trade intensity of below 15 percent.

Carbon black relies mostly on renewable and other sources of energy that are relatively less carbon intensive than coal. Nearly all of the energy consumed by nitrogenous fertilizers is from natural gas. In chemical sub-industries, a portion of the carbon contained in the energy source is also sequestered in the product rather than emitted to the atmosphere. Data on certain energy uses by type for sub-industries not shown in figure 16 is not publicly available and percentage shares could not be computed.

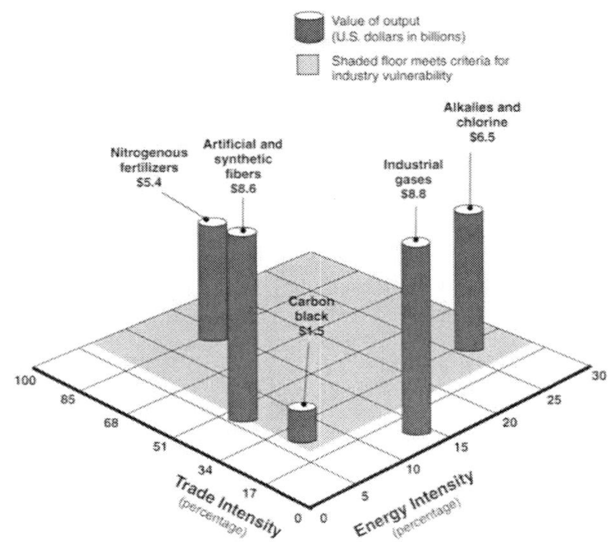

Source: GAO analysis of Department of Commerce energy data for 2006, output data for 2006 and 2007, and trade data for 2007.

Notes to figure 15

Figure 15. Energy and Trade Intensity Indicators Vary by Sub- Industry

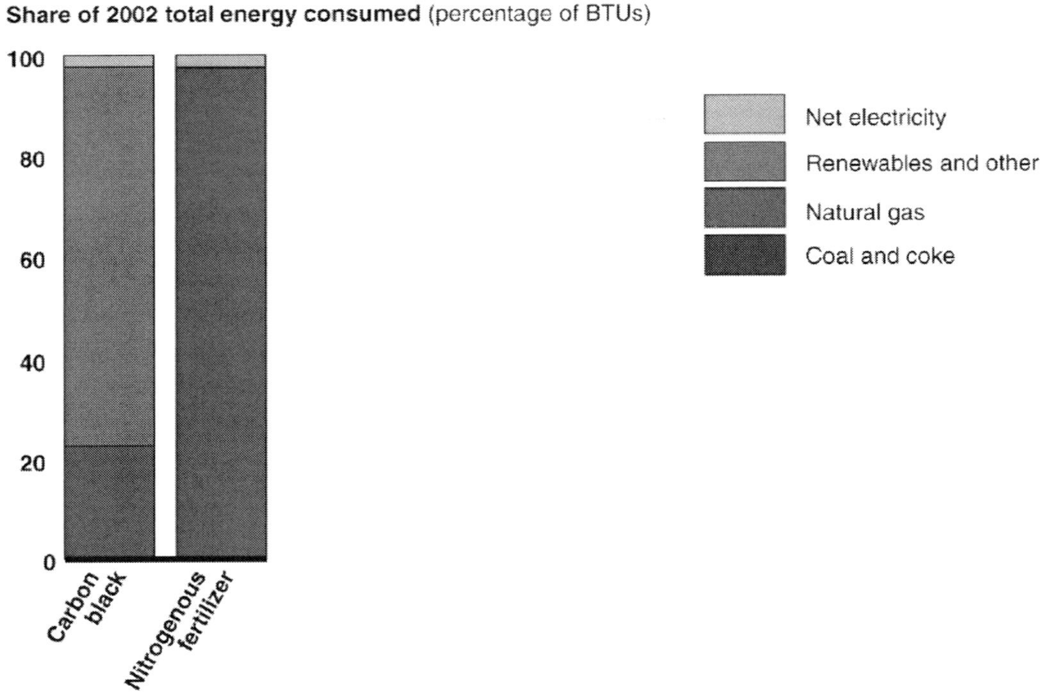

Source: GAO analysis of data from the Department of Energy.

Notes to figure 16

Figure 16. Type of Energy Used Is Important for Carbon Intensity

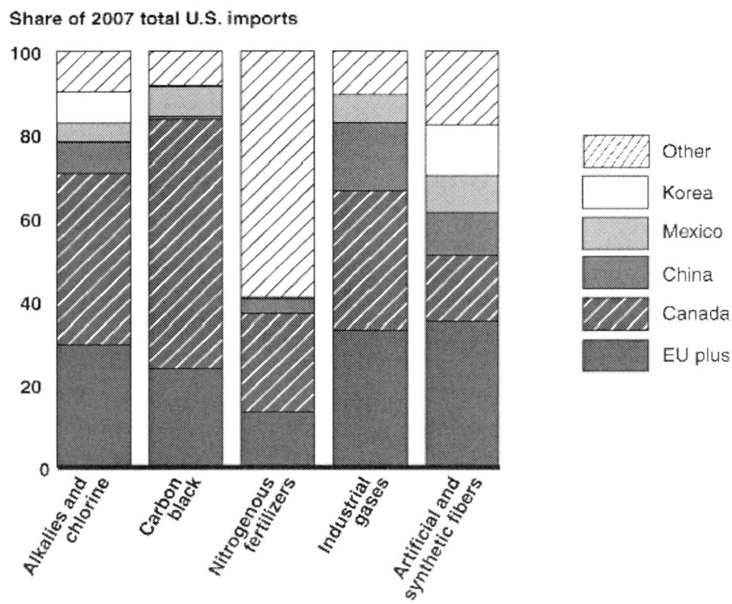

Source: GAO analysis of data from the Department of Commerce.

Notes to figure 17

Figure 17. Source of Imports Is Important for Trade Intensity

For the chemicals examples, total imports exceed total exports. Nitrogenous fertilizers is the most trade intensive, and less than 40 percent of imports are from Annex I countries—Canada and EU Plus— while 40 percent of imports are from either Trinidad and Tobago or countries in the Middle East. Conversely, imports from Canada provide 60 percent of imports for carbon black. Imports from China, Mexico, and Korea account for 20 percent or more of imports in industrial gases and artificial and synthetic fibers.

Alkalies and chlorine is the only sub- industry example where exports comprise more than 60 percent of trade flows, with 15 percent to Brazil, 13 percent to Canada, and 12 percent to Mexico.

From 2002 to 2007, the share of imports from China has grown for alkalies and chlorine, industrial gases, and artificial and synthetic fibers from a share of less than 2 percent to a share of between 8 percent and 16 percent. According to representatives from the chemical industry, growing imports from China are also evident for industry downstream products, such as plastics.

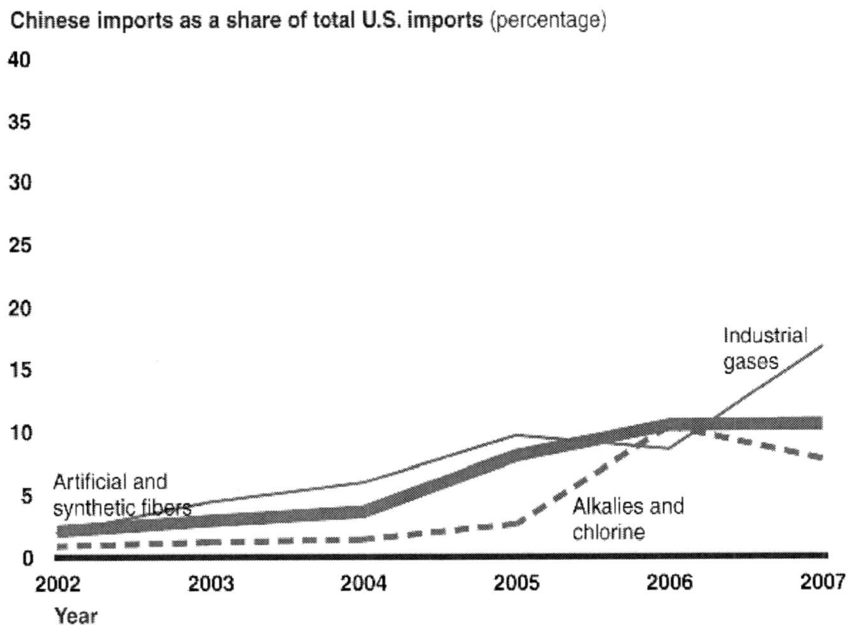

Total U.S. imports by value (U.S. dollars in millions)						
Year	2002	2003	2004	2005	2006	2007
Alkalies and chlorine	$160	207	253	454	460	400
Industrial gases	$125	128	148	160	160	176
Artifical and synthetic fibers	$1,626	1,644	1,814	2,230	2,322	2,471

Source: GAO analysis of data from the Department of Commerce.

Notes to figure 18

Figure 18. Sub-Industries With Growing Share of Imports from China

2.5. Summary of Industry Examples

This section provides data on energy and trade characteristics for sub-industries within primary metals, nonmetallic minerals, paper, and chemicals. As shown by the data, characteristics that contribute to an industry's or subindustry's potential vulnerability to adverse competitiveness effects vary significantly among the examples provided. Additional variation would likely exist for factors not discussed, such as transportation costs and access to markets for natural resources, capital, and labor. As a result, the data provided are not sufficient to determine if one sub-industry is more vulnerable than another. Instead, the data suggest that an assessment of vulnerability that is based only on a sub-industry's energy and trade intensity may mask important differences in vulnerability among industries assessed.

- **Primary metals examples**: Primary aluminum is the most energy and trade intensive, yet the largest share of imports is from Canada. Iron and steel is the largest sub-industry that is both energy and trade intensive and also has the greatest reliance on coal and coke energy sources. Steel manufacturing and aluminum products fall short of meeting the trade- intensity criteria, but compared with the other examples, have the largest share of imports from China and the most pronounced trend of an increased import share from China since 2002.
- **Nonmetallic mineral examples**: Cement is the most energy intensive and has a relatively greater reliance on coal and coke energies. Glass is the most trade intensive and has the largest share of imports from China. Both cement and glass show an increasing share of Chinese imports since 2002. Concrete, while not meeting the trade-intensity criteria, exports more than it imports, with almost one-third of exports destined for markets in Mexico and the Middle East.
- **Paper examples**: Paper mills is the largest sub-industry example, although most imports are from Annex I countries. Almost one-fifth of pulp mill imports is from Brazil, although pulp mills is the smallest example that meets both energy and trade intensity criteria. Paperboard containers do not meet either vulnerability criteria, but show an increasing share of imports from China since 2002.
- **Chemicals examples**: Nitrogenous fertilizers is the most trade intensive, and 40 percent of imports are from Trinidad and Tobago or countries in the Middle East. Alkalies and chlorine are the most energy intensive and exports account for 79 percent of trade flows, of which, over one-fourth are to markets in Brazil and Mexico. Industrial gases do not meet either vulnerability criterion, but compared with the other examples, have the largest share of imports from China and the most pronounced trend of an increased import share from China since 2002.

3. Trade Measures

Introduction

In this section, we discuss the use of trade measures and output-based rebates to address potential competitiveness and environmental effects of a domestic emissions pricing system.

Through trade measures, such as a border tax adjustment or a border allowance requirement, international competitors without comparable climate policies would pay for greenhouse gas emissions associated with their exports to the United States. Output-based rebates have been proposed as another type of measure to offset the costs of climate policy for sectors that are exposed to unregulated competition through payments based on a per-unit government rebate or tax credit, or a per-unit allocation of emissions allowances.

How a trade measure is ultimately designed will have important implications for the effectiveness of the overall measure in addressing industry and environmental effects. In terms of output based rebates, the extent and design of the rebates could help address the industry effects, but could also affect the costs to other industries.

In this section we provide explanations of key features of trade and output- based measures included in climate change legislation introduced between April 2007 and June 2009. We also provide information on a range of views regarding the value of trade and output-based measures as policy tools to address competitiveness and environmental concerns. Additionally, we provide information on the potential implementation challenges associated with implementing these measures.

3.1. Different Types

American Electric and Power Border Allowance Requirement

Some of the climate change bills introduced between April 2007 and June 2009 included provisions for a border allowance requirement. A border allowance requirement proposal was developed by the American Electric Power (AEP) and trade union representatives with the International Brotherhood of Electric Workers (IBEW). The AEP/IBEW proposal was introduced in July 2007 in S. 1766, 110TH Cong. (2007) and was also included in various other legislative proposals.

Trade measures have been proposed to address competitiveness and environmental effects associated with implementing a domestic emissions pricing system, such as a cap-and-trade system or a carbon tax system. Generally, in the context of a domestic emissions pricing system, trade measures would impose a cost or other requirement on energy-intensive imports from countries with weaker climate policies. Various types of trade measures exist for addressing competitiveness effects associated with a domestic emissions pricing system. Among the cap-and-trade and carbon tax legislative proposals we examined, border tax adjustments or border allowance requirements are the most frequently proposed type of trade measures.

Border Tax Adjustment

According to experts, a border tax adjustment is a levy applied by the federal government, on certain imported goods at the border. Key features of a border tax adjustment include the following:

- *Tax proportionate to the imports' embedded carbon*: Generally, the tax applied to the import would be based on the carbon emissions associated with the production of the imported good.

- *Tax equivalent to domestic compliance costs*: Generally, the federal government would charge imported goods the equivalent of what a domestic producer of a similar good would pay to comply with a domestic emissions pricing system.

Border Allowance Requirement

According to experts, a border allowance requirement is a measure that would require importers to purchase allowances from the federal government prior to importing goods into the United States. Generally, key features of a border allowance requirement include the following:

- *Establishes a separate pool of allowances*: In some of the cap-and-trade bills proposed in 2008, the federal government would establish a reserve of allowances separate from allowances established in the domestic capand-trade system. The border allowances that importers would be required to submit would come from this reserve allowance pool.
- *Comparable action determination*: Importers of goods from foreign countries that do not take "comparable action" to the United States to limit greenhouse gas emissions, would be required to submit allowances to accompany exports to the United States of the covered good.
- *Documentation at the border*: Prior to entering the United States, importers would have to purchase allowances from the federal government, and submit a written declaration at the border stating that the good is accompanied by the required number of allowances. Generally, the number of allowances required would be proportionate to the embedded carbon content of the import.

Supporters' Views

Some policy analysts, industry stakeholders, climate change experts, and government officials argue that trade measures help address competitiveness and environmental effects associated with domestic emissions pricing systems. For example, some supporters contend that trade measures may do the following:

- *Prevent a decline in output by U.S. producers*: Trade measures would help level the playing field by imposing similar costs on imports from countries without comparable carbon mitigation policies. The potential exists for certain U.S. firms to lose business to energy-intensive imports from countries with weaker climate policies. Moreover, firms could further lose international competitiveness due to the indirect costs of purchasing more expensive U.S. energy. As a result, U.S. firms in vulnerable industries could suffer a loss in output, profits, and employment.
- *Prevent carbon leakage*: Trade measures would prevent U.S. energy- intensive industries from shifting production to countries without comparable carbon mitigation policies, resulting in carbon leakage, where emission reductions in the United States are replaced to some degree by production and emission growth in less

regulated countries. Supporters of trade measures stated that carbon leakage could lead to higher volumes of greenhouse gas emissions worldwide.

- *Create leverage on other countries to reduce emissions:* Trade measures would create incentives for major trading partners to adopt similar climate policies. For example, proponents of trade measures argue that trade measures could potentially provide leverage to U.S. climate negotiators in their efforts to establish a global framework that includes other major emitting nations. Some industry stakeholders with whom we spoke, for example, stated that the potential benefit of a trade measure in increasing pressure on developing countries may trump its adverse economic effects. These stakeholders said that countries' statements of opposition to proposed trade measures provided evidence of their effectiveness in providing leverage.
- *Political acceptability*: If a climate change bill is to pass, it will be necessary that the bill include a provision like a trade measure to address competitiveness effects. Policy experts stated that most climate change proposals with trade measures have garnered support from industries that could be affected by a domestic emissions pricing system.

Opponents' Views

Some policy analysts, climate change experts, and government officials raise concerns that trade measures may motivate retaliatory actions, undermine efforts to secure multilateral consensus, and generate little leverage. For example, some opponents contend that trade measures may do the following:

- *Motivate retaliatory action*: An important implication of U.S. trade measures is how they would be perceived by other countries and impact negotiations aimed toward an international climate agreement. Opponents argued that trade measures could be viewed by other countries as an antagonistic step, and could lead to retaliatory action from other countries.
- *Undermine efforts to secure international consensus*: While there is general agreement that the global scope of the climate change problem will require actions from all major emitting nations, there are differing views on the impact that trade measures may have in securing an international consensus. Foreign government officials with whom we spoke, declined to give formal positions on trade measure proposals given that a climate change bill has not yet passed Congress. But some officials expressed concerns over the unilateral application of a trade measure by the United States and said that competitiveness effects would be better addressed through multilateral discussions. Because multilateral approaches involve commitments to reduce greenhouse gas emissions by more than one country, they also level the playing field. Multilateral approaches include international agreements, such as the Kyoto Protocol, or international action, as when countries independently decide to reduce their emissions of greenhouse gases. Multilateral approaches may involve sectoral agreements, leveling the playing field worldwide within an industry. Foreign embassy officials also stated that trade measures are not likely to be effective

tool to leverage other countries to take steps to reduce emissions and that a unilateral U.S. trade measure could add to the challenge of developing multilateral consensus on climate change mitigation efforts.

- *Generate little leverage on other countries to reduce emissions:* The effectiveness of trade measures in creating leverage on foreign countries to reduce emissions will vary with the share of their output that is imported into the United States compared with the share that is consumed domestically or exported to other countries. In addition, some policy experts we spoke to note that in some cases, the emission costs that foreign firms will have to pay on their imports into the United States may not be sufficient to motivate them to change domestic production processes. For example, in each of the vulnerable industries listed in table 2, less than 1 percent of total Chinese output is imported into the United States, although the share may be higher for individual firms. Instead, most of the Chinese output in these industries is consumed domestically or in countries that may or may not impose a domestic emissions pricing system. Nonetheless, in certain cases, a U.S. trade measure may be effective in providing some degree of leverage. For example, U.S. iron and steel imports represent more than 8 percent of output in Brazil and more than 10 percent of output in Mexico. In addition, U.S. imports of nitrogenous fertilizers from the Middle East are more than 20 percent of production from that region.

Implementation Challenges

Trade measures could present implementation challenges, particularly with obtaining necessary data to measure the carbon content of imports, and to evaluate and assess climate change actions of other countries. Examples of potential implementation challenges include the following:

- *Determining the embedded carbon content in imports*: Generally, imposing a trade measure on an import would require a determination of the embedded carbon content, or the amount of greenhouse gases emitted during the production of the imported good. According to experts, accurately measuring the embedded carbon content for specific items would be challenging. Furthermore, climate policy experts point out that while determining the embedded carbon content in standardized products like steel, primary aluminum, and basic chemicals would be difficult, it would be even more challenging to assess the embedded carbon content of products that rely on these goods for final assembly.

  Additionally, climate policy experts noted that, variations in the type of energy used can result in different carbon intensities for goods that appear identical at the border. To accurately assess the embedded carbon content of a product, the administrator of the program would be required to obtain specific plant-level information on the production process of the import from foreign countries.

Table 2. U.S. Imports as a Share of Foreign Output (percentage of metric tons)

	Iron and steel	Primary aluminum	Cement	Pulp products	Nitrogenous fertilizers
Brazil	8.3	5.0	1.1	10.8	3.5
China	0.5	0.3	0.9	0.1	0.5
India	0.7	0.0	0.0	n/a	0.0
Mexico	10.6	n/a	5.6	1.1	8.3
Middle East	0.1	6.5	0.2	0.0	22.1

Source: GAO analysis of production data from the U.S. Geological Survey and the Food and Agriculture Organization and trade data from the Department of Commerce.

Notes to Table

Data are for 2007, except for cement and nitrogenous fertilizers with data for 2006.

Iron and steel production figures include those for pig iron, direct-reduced iron, and raw steel. Cement production figures are for hydraulic cement.

Pulp products production figures represent the total for "pulp for paper" products within FAOSTAT. Nitrogenous fertilizer production figures represent tonnage by total nutrients within FAOSTAT. Not included in the iron and steel import figures above were about 38 million liters from Mexico.

Accurately measuring the embedded carbon content of an import would require significant data, resulting in several data reliability concerns with obtaining carbon data from international inventories. Experts and industry stakeholders alike reported that obtaining data from foreign producers could be a challenge, and that using data from international inventories could raise several data reliability concerns as the United States would have no way of verifying that the data obtained was accurate. In addition, under the S. 3036 110TH Cong. (2008), importers would have to submit a declaration statement at the border declaring the imported good is accompanied by the required number of allowances. Agency officials we spoke to noted that officials would not know if the information contained in the declaration was accurate.

- *Evaluating and assessing climate change actions*: As previously discussed, a key feature of a trade measure provision is the requirement of determining the comparability of other countries' actions to address climate change. According to policy experts, among some of the cap-and-trade bills introduced, some have not clearly defined or assessed the method for determining how other countries' actions on climate change would be assess by the U.S. government. In order for the implementing agency to assess the comparability of different countries, the criteria for making this determination would have to be defined either in the legislation or implementing regulations.

Design Trade-Offs

In designing a trade measure, there are a range of possible options to consider as illustrated in Table 3. The design features of trade measures involve different trade-offs that will have implications for the effectiveness of the overall measure in addressing competitiveness and environmental concerns.

Table 3. Policy Experts Cite Trade-offs Involved in the Design of Trade Measures

Feature	Options	Design trade-offs
Date when the trade measure goes into effect	The trade measure can go into effect simultaneously when a domestic cap-and-trade system goes into effect or at a later date.	Having the trade measure go into effect simultaneously when the domestic cap-and-trade system goes into effect may reduce impacts on U.S. industry competitiveness. Delaying the implementation of the trade measure allows time for international climate negotiations and for countries to take action on reducing their emissions.
Type of products covered	Climate bills in the 110th Congress limited coverage of the trade measure to specified products such as importers of steel, aluminum, and cement; whereas other bills expanded coverage to manufactured goods that met eligibility criteria.	Limiting product coverage may exclude industries vulnerable to competitiveness effects Expanding product coverage increases federal challenges to track emissions
Defining comparable action	Proposals have varied in stringency and level of detail in outlining the methodology for determining comparable action.	Quantitative criteria for assessing comparable action may not measure the level of effort on climate policy that a country actually undertakes. Flexibility and discretion on comparability determinations increases uncertainty for industry stakeholders.
Calculating border allowance requirements	Proposals have varied in the level of detail in outlining the methodology for calculating the number of allowances industries will be required to submit at the border. Earlier proposals based border allowance requirements on averages for country industry sectors; H.R. 2454 delegated the decision to the future implementing agency.	Basing border allowance requirements on industry average increases implementation feasibility as firm-level data likely not available. Using industry average for baseline may limit incentives to improve efficiency and would penalize efficient firms.

Source: GAO analysis of selected climate change legislation introduced between April 2007 and June 2009 that included trade measures.

For example, a key design consideration is the timing for when the trade measure would go into effect. Delaying the implementation of the trade measure could result in the measure being a more effective leverage tool because it would provide other countries with an opportunity to implement similar carbon mitigation policies. On the other hand, industry stakeholders with whom we spoke argued that a delay in implementing the trade measure could adversely affect U.S. industries by incurring costs under a domestic emissions pricing system that their foreign competitors would not face. According to some stakeholders, this could result in U.S. industries being at a competitive disadvantage compared to foreign competitors.

The method for determining the comparability of other countries' actions also involves important trade-offs. Given the different approaches being utilized to address climate change, questions have been raised over whether these different actions would be deemed comparable to U.S. efforts. Industry groups have argued that the method for determining comparability should be based on quantitative criteria that are equivalent to criteria U.S. producers must meet. However, providing flexibility on determining comparability allows countries to utilize the approach best suited for reducing emissions in their individual country.

3.2. Alternative: Output-Based Rebates

Output-based rebates have been proposed as policy alternatives to using trade measures. These rebates would be provided to energy-intensive (or greenhouse gas-intensive) and trade-exposed industries to cover their increased costs from the carbon pricing system. Under legislative proposals, EPA would determine the average energy (or greenhouse gas) intensity per unit of production for each relevant industrial sector and then distribute allowances based on each facility's amount of production. Although some industry stakeholders note that output-based rebates address competitiveness effects by compensating covered firms for incurred costs, some climate policy experts note that output-based rebates could distort pricing, could reduce incentives for firms to engage in conservation, and may drive up the cost of the program.

Output-based Rebates

Output-based rebates are measures designed to financially rebate industries for the costs incurred under a domestic emissions pricing system. According to policy experts, key features of an output-based rebate include the following:

- *Tied to firm's level of output*: Generally, output-based rebates would be tied to a firm's level of output. For example, firms that expand their operations will receive a larger rebate, while firms that downsize but continue to produce in the United States will receive a smaller rebate. Firms that move offshore or shut down would receive no rebate. A rebate can take the form of a per-unit rebate or tax credit, or a per-unit allocation of emissions allowances.
- *Compensates for incurred costs*: For a facility with an average level of energy intensity (or greenhouse gas intensity) these rebates would cover both direct and indirect costs. These would include their own emissions and costs associated with higher energy prices and technology purchases to improve energy efficiency. Providing allowances based on average industry intensity, energy intensity and production levels, rather than providing them based on each firm's historical

emissions, is intended to reward the most efficient facilities and create an incentive for continued efficiency improvements.

H.R. 2454 111th Cong. (2009) included provisions for output-based rebates to offset the costs incurred by firms to comply with a cap-and-trade system. Two other climate change bills, H.R. 7146, 110th Cong. (2008) and H.R. 1759, 111th Cong. (2009), also incorporated output-based rebates. In all three proposals, allowances would be distributed among certain energy- and trade- intensive industries according to an individual firm's output.

Supporters' Views

Some policy analysts, climate change experts, and government officials argue that output-based rebates may address some of the limitations associated with using trade measures. For example, some supporters contend that output- based rebates may do the following:

- *Mitigate costs incurred by U.S. producers*: Output-based rebates are necessary to compensate U.S. firms for costs incurred while complying with a domestic emissions pricing system. Supporters note that trade measures only apply costs to foreign competitors at the border, while output-based rebates affect producers' costs for both domestic and export markets. Among the key supporters of output-based rebates are stakeholders in energy-intensive industries who note that energy costs are a substantial portion of their manufacturing costs. According to these stakeholders, proposals that mitigate costs at the production level, either by allocating free allowances to qualifying facilities or rebating the costs of allowances offsets the disincentive created by higher energy prices in the United States.
- *Reward efficient firms*: Output-based rebates are generally allocated in proportion to current levels of production rather than being fixed to historical measures. Supporters argue that because the rebates or allocations are tied to production levels, individual firms have an incentive to increase their production. According to supporters, rebates reward the most productive plants and stimulate investments in efficient technology. Moreover, the rebates would benefit firms that face competition from foreign suppliers in markets at home or global export markets or both.
- *Use U.S. data*: Implementing output-based rebates may be more manageable than implementing trade measures. For example, in proposed legislation, agencies would only require data from U.S. data sources to identify firms that would be eligible for output-based rebates. This differs from trade measures which would require data from foreign firms' production, which could be more difficult to obtain. In addition, legislative proposals have generally been explicit about the U.S. data sources and criteria to be used in identifying eligible industries.

Opponents' Views

Some policy analysts, climate change experts, and government officials argue that despite several advantages identified with using output-based rebates, several potential limitations also exist. For example, some opponents contend that output-based rebates offset the costs of climate policy for sectors that are exposed to unregulated competition, and thereby lower the incentives for those firms to engage in conservation and reduce their energy intensity.

Similarly, output-based rebates would likely result in relatively lower output prices for the products produced by firms receiving rebates, thereby reducing the incentives for consumers of those products to conserve. Moreover, some climate policy experts argue that output-based rebates create a subsidy for certain energy-intensive industries which would require non energy-intensive firms to engage in greater efforts to reduce the emissions intensity of their production.

Potential Implementation Challenges

Although the data requirements for an output-based measure may be more manageable than those of a trade measure, some challenges related to identifying electricity distribution data exist. For example, under HR 2454, 111th Cong. (2009), EPA would be required to obtain production data on various commodities from energy-intensive industries, including electricity distribution data. Data on electricity use on a facility-by-facility basis is not currently collected by government agencies. Provisions in either the legislation or in implementing regulations would have to be explicit on how to obtain data on electricity use.

Additionally, according to EPA officials, the methodology outlined in HR 2454, 111th Cong. (2009), for how to determine eligible industries does not align with data that would be collected under EPA's proposed mandatory greenhouse gas reporting rule. According to climate policy experts, standard mechanisms for reviewing eligibility for the rebates are needed to ensure that the policy does not over compensate certain industries.

3.3. Legislative History of Trade Measures and Output-Based Rebates

Between April 2007 and June 2009, several cap-and-trade and carbon tax bills were introduced with provisions to address the potential competitiveness and environmental effects of implementing a domestic emissions pricing system. For example, of the cap-and-trade bills that we examined and that were introduced during that period, some included trade measures such as an allowance requirement at the border or a border tax.

As illustrated in figure 19, two bills introduced in 2009 have included provisions for using output-based rebates. The most recent bill, H.R. 2454, 111th Cong. (2009) includes a provision for output-based rebates. Under that bill, energy intensive and trade exposed industries would receive allowances to cover their increased costs.

Several industry stakeholders with whom we spoke, reported that output-based rebates and trade measures could function as complimentary measures to address competitiveness concerns. For example, in H.R. 2454, 111th Cong. (2009), the output-based rebates would be phased out by 2035 unless the President decides to extend them. In that bill, the trade measure would go into effect after 2020, unless the President determines it would not be in the national interest and Congress passes an affirmative joint resolution within 90 days.

Some stakeholders have stated that having the trade measure to go into effect at a later time may provide time for the United States to be part of international negotiations on climate change.

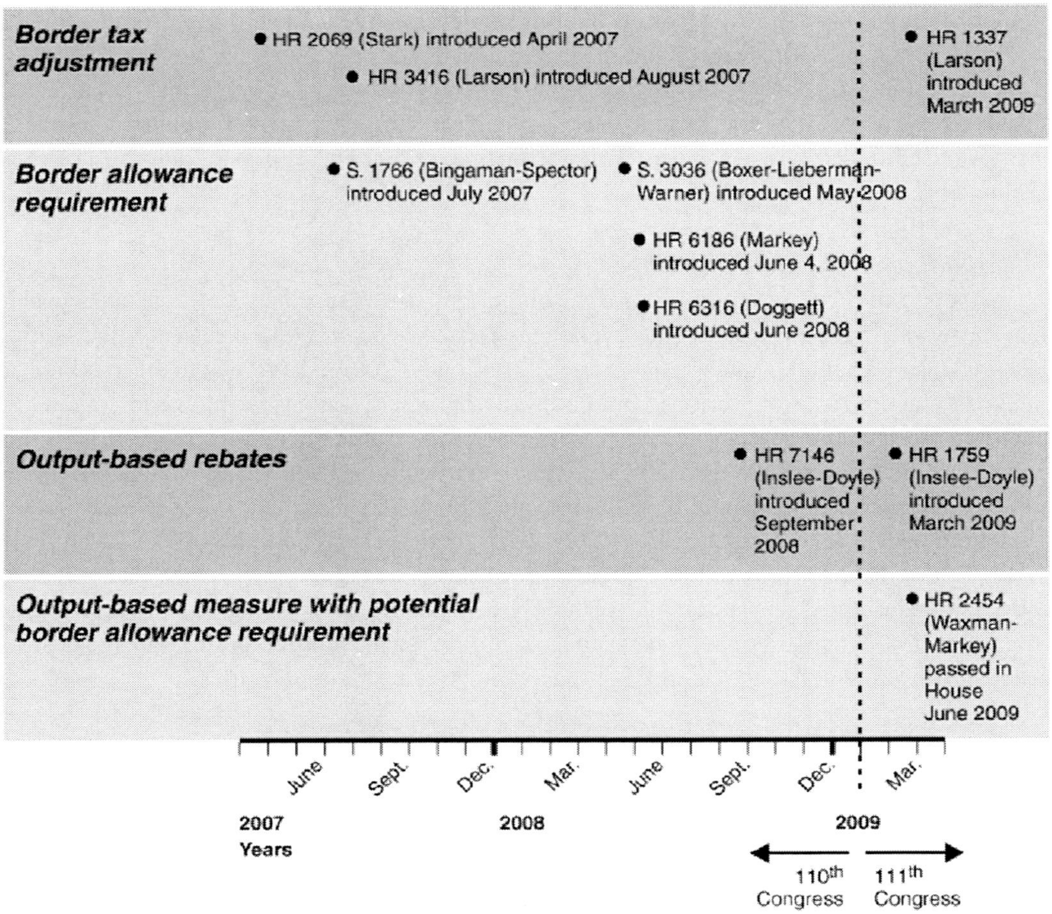

Source: GAO analysis.

Notes to figure 19

Figure 19. Proposals Move toward Using Output-Based Rebates to Address Competitiveness Concerns

4. International Trade Implications

Overview

Introduction

In this section, we discuss potential international implications of a U.S. trade measure, such as concerns about countries possibly retaliating in response to a trade measure or its potential impact on the multilateral trading system. In addition, we provide information on the WTO dispute settlement process, a forum that facilitates the resolution of specific trade disputes. We provide information on the WTO provisions and key questions that may apply if a U.S. trade measure or an output-based rebate were to be challenged at the WTO.

Assessing the international trade implications is difficult for a number of reasons. One is that it depends in part upon how other nations reduce their carbon emissions, and whether they perceive any U.S. measures as likely to affect their exports. In addition, the outcome of a WTO challenge, if any, would be uncertain and may depend on how the measure is implemented.

Potential Bilateral and Multilateral Trade Impacts

According to experts and foreign officials we spoke with, U.S. climate change trade measures may have implications for U.S. and multilateral trade relations.

- *Potential for bilateral trade retaliation:* Countries may view U.S. trade measures as trade restrictions or sanctions, which could lead them to implement restrictions against U.S. exports. Other countries could potentially develop counter measures based on a different test of "comparability," such as historical or per capita emissions, according to trade experts we spoke with.
  Although no other country has yet to implement a trade measure based on greenhouse gas emissions, European officials have previously considered the use of trade measures on countries, including the United States, that have not taken actions that match European standards. In addition, support within other countries for imposing trade measures could increase if the United States implements a trade measure.
- *Potential for increased trade tensions:* A broader concern is that escalating retaliation between the United States and other countries could lead to significant global trade tensions. Given the volume of trade in goods that could potentially be affected by trade measures linked to greenhouse gas emissions, some officials and experts have argued that escalating tensions and responses to these measures could ultimately do significant harm to the functioning of the multilateral trading system under the WTO. Trade experts told us that trade measures and responses could potentially pose systemic challenges to the WTO and its dispute resolution system. For example, multiple challenges brought up by other countries based on different imported products could tie up the system.

4.1.WTO Compliance Questions

WTO Dispute Settlement Process

Consultation (up to 60 days)

Panel Review (generally up to 6 months, but in no case more than 9 months)

Appellate Stage (60 to 90 days)

DSB Decision (generally up to 9 months and up to 12 months if the panel report is appealed)

Implementation Stage (if immediate compliance is impractical, member given a 'reasonable period of time' to comply, which normally should not exceed 15 months from adoption of report up to 15 months)

Table 4. Key Questions That May be Considered in a WTO Challenge

Key questions	Discussion	WTO provisions
1. Is the trade measure consistent with WTO market access requirements?	Customs duties or other duties or charges on imports in excess of schedules of concessions may not be imposed. Subject to exceptions, prohibitions or restrictions, other than duties, taxes or other charges, on imports may not be imposed.	GATT Article II, XI
2. Is the trade measure consistent with WTO non-discrimination requirements?	Measure must treat imported products no less favorably than like domestic products.	GATT Article III
	Measure must also not discriminate among goods from different countries.	GATT Article I
3. If not WTO consistent, is the trade measure covered by a WTO environmental exception?	Measure relating to conservation of exhaustible natural resources or that is necessary to protect human, animal or plant health is covered under WTO exceptions.	GATT Article XX (b), (g)
	To be covered, measure must not be applied in a manner which would constitute a means of arbitrary or unjustifiable discrimination or be a disguised restriction on international trade	GATT Article XX chapeau
4. Are output-based rebates consistent with WTO rules governing subsidies?	The rebates must not involve a specific subsidy or other benefit that causes adverse effects to other WTO Members.	ASCM

Source: GAO analysis of legal scholars' views on questions and provisions that may be relevant in analyzing trade measures contained in climate change legislation for consistency with WTO rules.

A unilateral U.S. trade measure or system of rebates linked to a climate change policy may be challenged by other countries through the WTO's dispute resolution process. Countries could potentially raise different types of challenges depending on the design of the trade measure or rebate, and several key questions may be relevant in the consideration of a WTO dispute. These questions stem from the WTO provisions under which the trade measures or rebates may be challenged and reviewed, including the General Agreement on Tariffs and Trade (GATT), and the Agreement on Subsidies and Countervailing Duties (ASCM).

The following pages describe in more detail the questions, factors, and WTO provisions listed in table 4.

WTO Dispute Settlement Process

The WTO's dispute settlement system facilitates the resolution of specific trade disputes and serves as a vehicle for upholding trade rules and preserving the rights and obligations of WTO Members under WTO agreements. WTO Members may request consultations concerning measures affecting the operation of such WTO agreements. After consultations, WTO Members can request the establishment of a panel to hear their claims regarding alleged inconsistencies of other Member's measures. The panel will issue a report of its findings, which will be adopted by the WTO Dispute Settlement Body (DSB), unless a party to the dispute appeals the panel's report, which is reviewed by the WTO Appellate Body. WTO disputes can take several years or more to complete the entire process. If the responding party does not prevail and fails to comply with the rulings and recommendations of the DSB, the complaining party may seek authority to suspend concessions or other obligations under the WTO agreements.

WTO Compliance

The following observations are based on comments from various legal scholars and other non-governmental commentators on questions and provisions that may be relevant in analyzing trade measures contained in climate change legislation for consistency with WTO rules.

1. Is the Trade Measure Consistent with WTO Market Access Rules Commitments?

An initial question involves how a trade measure would be interpreted and the corresponding GATT article under which the measure would be reviewed.

- GATT Article II:1(b) prohibits a WTO Member from applying customs duties or other duties or charges to imports in excess of the Member's schedule of tariff concessions.
- GATT Article XI: 1 prohibits prohibitions or restrictions, other than duties, taxes or other charges, on imports, with certain exceptions.
- According to legal experts, a "border tax" imposed on imports equivalent to a tax imposed on like domestic products may be WTO compliant.

2. Is the Trade Measure Consistent with WTO Non-Discrimination Requirements?

Trade measures must also be consistent with WTO non-discrimination provisions covering most favored nation status treatment and national treatment.

- *Does a trade measure treat two like products differently depending on country of origin?* GATT Article I:1 provides, in part, that with respect to customs duties and charges imposed on or in connection with importation and exportation, and the method of levying such duties and charges, any advantage granted by any WTO Member to any product originating in any other Member is to be accorded immediately and unconditionally to the like product originating in or destined for all other WTO Members.
- *Does a trade measure treat imported products less favorably than like domestic products?* GATT Article III:4 provides, in part, that imported products shall be accorded treatment no less favorable than that accorded to like products of national origin in respect of all laws, regulations, and requirements affecting their internal sale, offering for sale, purchase, transportation, distribution, or use.

Shrimp-Turtle Case - In this 1996 case, four WTO members challenged as discriminatory a U.S. embargo on the importation of certain shrimp and shrimp products from countries that had not been certified by the Department of State as having a comprehensive sea turtle conservation regime. The United States argued that the measure was justified under article XX (g).

However, the WTO panel found, in part, that the U.S. measure was not justified under article XX (g) because it met the panel's test of being a "risk to the multilateral trading system." Upon appeal, the Appellate Body rejected the panel's test and found that the measure fell within article XX (g). However, the Appellate Body also found the U.S. measure had been applied in an arbitrary and discriminatory manner.

The United States was ultimately able to comply with WTO rulings by making changes in the way the law was administered, including revising guidelines to establish greater transparency and due process in country certification decisions.

3. If the Trade Measure Is Not Permissible Under Core WTO Obligations, Would it Be Covered Under an Environmental Exception?

If a trade measure does not comply with GATT market access or nondiscrimination provisions, it may still be justified under certain GATT environmental exceptions. A key consideration would be the interpretation of the intended purpose of the measure and whether it is designed to serve an environmental objective.

- *Is the trade measure designed to serve an environmental objective?*
  - GATT Article XX (b) provides an exception for measures necessary to protect human, animal, or plant life or health.
  - GATT Article XX (g) provides an exception for measures relating to the conservation of exhaustible natural resources if such measures are made effective in conjunction with restrictions on domestic production or consumption.
- *How is the measure applied?* The introduction (chapeau) to GATT Article XX states that such measures must not be applied in a manner that would constitute a means of arbitrary or unjustifiable discrimination or a disguised restriction on international trade.

4. Are Output-Based Rebates Consistent with WTO Rules Governing Subsidies?

The ASCM imposes disciplines on certain kinds of subsidies to domestic producers, such as those that cause adverse effects to foreign competitors.

- *Is an output-based rebate or distribution of free allowances a subsidy?* To be considered a subsidy under the ASCM, a system of rebates would need to involve (1) a "financial contribution" by the government or any form of income or price support that (2) confers a "benefit" to the recipient and (3) is "specific."
- *Is the rebate an "actionable" subsidy that causes "adverse effects" to the interests of other WTO Members?* A further key question is whether a rebate, if found to be a subsidy, would cause "serious prejudice" to the interests of other WTO Members.

APPENDIX 1

Objectives, Scope, and Methodology

Our research objectives were to examine: (1) information on estimating industry effects; (2) examples of industries that may be vulnerable to a loss in international competitiveness from emissions pricing; (3) trade measures and other approaches to address competitiveness issues; and (4) potential international implications of trade measures.

To address all of these objectives, we (1) reviewed relevant climate change academic literature and federal agency documents; (2) reviewed and analyzed studies and data on climate change from a variety of sources, including past GAO work, the U.S. Census Bureau and Departments of Commerce and Energy, international organizations, policy institutes, and universities; (3) interviewed agency officials from the Office of the U.S. Trade Representative (USTR); the Environmental Protection Agency (EPA); the Departments of Commerce, State, Energy, and Treasury, and the Department of Homeland Security's Customs and Border Protection (CBP); (4) conducted interviews with subject-matter experts selected from a population of individuals from government, academia, business, and professional organizations. Several criteria were used for selecting experts to interview including: type and depth of experience, the expert's recognition in the professional community, relevance of published work, professional affiliations, present and past positions held, and other subject matter experts' recommendations.

Objective 1 and 2

To better understand how implementing a domestic emissions pricing system would likely affect the international competitiveness of U.S industries, we reviewed relevant studies from GAO, EPA, the Department of Energy's Energy Information Agency (EIA), and international organizations. We also reviewed documents and interviewed experts from policy institutes, universities, and industry.

Recently proposed legislation identifies industries potentially vulnerable to adverse competitiveness effects as those with: (a) either energy intensity or greenhouse gas intensity of 5 percent or greater; and (b) a trade intensity of 15 percent or greater.[4] Since calculation of greenhouse gas intensity requires an emissions price—not yet determined—and detailed industry data on greenhouse gas emissions—not currently available—we focus our analysis on industry energy intensity. However, energy intensity is also an imperfect measure of potential vulnerability since energy use is not the only source of industrial greenhouse gas emissions, and an industry's energy expenditures may rise with to a change in market prices, even if the industry's greenhouse emissions decline. Additionally, many factors may determine the vulnerability of any particular firm, including transportation costs, supply chain relationships, and access to markets for natural resources, capital, or labor.

Based on industry and expert studies, most industries that meet both vulnerability criteria are within primary metals, nonmetallic minerals, paper products, and chemicals.[5] Quantitative analysis of the estimated competitiveness impacts of U.S. greenhouse gas emissions pricing is relatively limited. We present results for two studies that separate the international competitiveness effects from U.S. emissions pricing from the domestic market effects. Estimated results from these two studies include findings that are broadly consistent with those of other studies. For example, a study of climate policy impacts on U.S. industries was commissioned by the National Committee on Energy Policy and studies of the European Union's climate policies have been performed for the European Commission and the International Energy Agency. In each of these studies, the list of vulnerable industries is generally consistent, as is the finding that estimated impacts could be greater for energy-intensive industries.

To illustrate variation in economic characteristics among these four manufacturing industries, we analyzed: (1) energy consumption data from the U.S. Census Bureau's Annual Survey of Manufacturers; (2) energy use data from EIA's Manufacturing Energy Consumption Survey; (3) international trade data from the Department of Commerce's International Trade Administration; and (4) U.S. and international production data from the Department of Commerce's Bureau of Economic Analysis, the U.S. Geological Survey and the United Nations Food and Agricultural Organization (FAO). For analysis of sub-industries, we examined data with a Census Bureau North American Industry Classification System (NAICS) code of 5 to 6 digits, depending on how the data were reported in each of the databases reviewed and for consistency of comparison with other studies. For the vulnerable sub- industry charts, we applied multiple criteria to show variation in industry characteristics of energy intensity, trade intensity, and primary trading partners. For example, we selected sub-industries that met both the energy and trade intensity criteria, examples that met only one criterion, and examples that met neither, but had significant imports from non-Annex I countries. We do not identify the complete list of potentially vulnerable subindustries, a list that will vary depending on the level of aggregation in product lines that is examined. The specific list of example sub-industries we selected, along with their corresponding NAICS codes can be found in table 5.

We tested the data we analyzed for internal consistency and for consistency with other key published studies, interviewed agency and industry officials regarding appropriate use of the data, and reviewed source data information regarding the entity's methodology and actions taken to ensure data reliability. We determined that the data were sufficiently reliable for our use. We note, however, that industry characteristics may change over time and may vary for firms within each sub-industry selected. To supplement our data analysis, we conducted interviews with industry groups such as steel, aluminum, chemicals, and the Energy-Intensive Manufacturer's Working Group.

Objective 3

To identify and compare key features of proposed trade measures, we reviewed and analyzed selected climate change legislation introduced between 2007 and 2009, and congressional hearing records. From these documents we extracted information regarding features of the trade measures, such as objectives, scope of coverage, effective date for trade measures, and other key features. To identify how features of trade measures could address the potential economic and environmental effects of emissions pricing, we reviewed studies and reports obtained from our literature review, including past GAO reports on climate change and studies published by climate change policy institutes such as Resources for the Future, the Peterson Institute, and the Pew Center on Global Climate change. We also interviewed subject matter experts and agency officials from USTR; EPA; the Departments of Commerce, State, Energy, and Treasury, and CBP, regarding these issues and to identify the potential implementation and administrative challenges with using trade measures. In our interviews with agency officials we discussed steps agencies have taken to anticipate implementation challenges. To obtain information about climate change policies in other countries and to learn about the potential impact of U.S. trade measures on bilateral relations and international negotiations, we also interviewed officials from the embassies of Australia, Brazil, Canada, China, European Union, Mexico, and Singapore.

Objective 4

To assess potential international implications of using trade measures, we interviewed subject matter experts, agency officials in EPA, CBP, USTR, EIA, and the Departments of Treasury, State, and Commerce, and foreign embassy officials. To discuss questions and provisions that may be relevant in analyzing trade measures and output-based rebates for consistency with WTO rules, we reviewed WTO documents, obtained information from our interviews with subject-matter experts and agency officials, and reviewed information obtained from our literature review.

Table 5. Sub-Industries Selected as Vulnerable Industry Examples

Sub-Industry Examples	North American Industry Classification System Codes
Primary Metals	331
Iron and steel mills	331111
Electrometallurgical products	331112
Steel manufacturing	33121, 33122
Ferrous metal foundries	33151
Primary aluminum	331312
Aluminum products	331314, 331316
Nonmetallic Minerals	**327**
Cement	32731
Concrete	32732
Lime and gypsum	3274
Glass	32721
Clay building material	32712
Mineral wool	327993
Paper Products	**322**
Pulp mills	32211
Paper mills	32212
Paperboard mills	32213
Paperboard containers	32221
Chemicals	**325**
Alkalies and chlorine	325181
Carbon black	325182
Nitrogenous fertilizers	325311
Industrial gases	32512
Artificial and synthetic fibers	32522

Source: U.S. Census Bureau, http://www.census.gov/eos/www/naics/

USTR provided technical comments on this report.

We conducted our work from October 2008 to July 2009 in accordance with all sections of GAO's Quality Assurance Framework that are relevant to our objectives. The framework requires that we plan and perform the engagement to obtain sufficient and appropriate evidence to meet our stated objectives and to discuss any limitations in our work. We believe that the information and data obtained, and the analysis conducted, provide a reasonable basis for any findings and conclusions in this product.

APPENDIX 2

Notes to Industry Figures

Notes to Figure 2: Energy intensity is calculated as the value of purchased fuels and electricity as a share of the value of output. Trade intensity is calculated as the value of total trade (exports plus imports) divided by the value of output plus imports. Data for value of output are from the Bureau of Economic Analysis Gross Domestic Product (GDP) by industry accounts, shipments by industry in current dollars.

Notes to Figure 3: Due to aggregation in source data, energy intensity for electro metallurgical products is assumed to be the same as iron and steel mills. Energy intensity data for primary aluminum is based on 2005 data. Aluminum products include industries characterized by the North American Industry Classification System (NAICS) with codes 331314 and 331316 and steel manufacturing industries represent codes 33121 and 33122.

Notes to Figure 4: Energy types characterized as "Renewables and other" includes residual and distillate fuel oil, liquefied petroleum gases, natural gas liquids, net steam (the sum of purchases, generation from renewables, and net transfers), and other energy that respondents indicated was used to produce heat and power or as feedstock/raw material inputs. Net electricity is the sum of purchases, transfers in, and generation from noncombustible renewable resources minus quantities sold and transferred out. Due to aggregation in source data, "Aluminum" includes primary aluminum and aluminum products.

Notes to Figure 5: "EU Plus" includes countries from the EU and other Annex I countries not represented elsewhere. Other than China, the largest two sources of imports for electrometallurgical products are the Republic of South Africa (17 percent) and Trinidad and Tobago (10 percent).

Notes to Figure 6: The current value of imports from all countries is shown in the table below the chart.

Notes to Figure 7: Energy intensity data for clay building materials are based on 2005 data because 2006 data were not available. Glass products include industries characterized by the North American Industry Classification System (NAICS) with codes 32721.

Notes to Figure 8: Energy use by type data for lime products excludes gypsum. Data on energy use by nonmetallic mineral sub-industry examples excluded from the chart are not currently reported by the Department of Energy.

Notes to Figure 9: After Canada and China, the two largest sources of imports for cement are Korea (9 percent) and Colombia (8 percent).

Notes to Figure 10: The current value of imports from all countries is shown in the table below the chart.

Notes to Figure 11: Due to aggregation in Bureau of Economic Analysis value of output data for paper mills with paperboard mills, value of output data for these two sub-industries are for 2006 and are from the Census Bureau's Annual Survey of Manufacturers, shipments by industry in current dollars. Aggregated data for 2006 are consistent between the two sources.

Notes to Figure 12: Data for coal and coke usage by pulp mills are not publicly available to avoid disclosure of individual firm information, such that percentage shares could not be computed. Data on energy use by paperboard containers is not currently reported by the Department of Energy.

Notes to Figure 14: The current value of imports from all countries is shown in the table below the chart.

Notes to Figure 15: Energy intensity data for alkalies and chlorine and carbon black is based on 2005 data because 2006 data is currently not available.

Notes to Figure 16: Data for certain energy uses by type for alkalies and chlorine, industrial gases, and artificial and synthetic fibers is not publicly available to avoid disclosure of individual firm information, such that percentage shares could not be computed.

Notes to Figure 17: Other than Canada, the two largest sources of imports for nitrogenous fertilizers are Trinidad and Tobago (26 percent) and the Middle East (14 percent), with countries in the Middle East defined by the U.S. Department of State.

Notes to Figure 18: The current value of imports from all countries is shown in the table below the chart.

Notes to Figure 19: In addition to these bills, Representatives John Dingell and Rick Boucher released a discussion draft on October 7, 2008. The discussion draft included a cap-and-trade proposal, and presents four options for allocating allowances under a cap-and-trade system.

APPENDIX 3

Works Cited

Aldy, Joseph E. & Pizer, William A. (May, 2009). "*The Competitiveness Impacts of Climate Change Policies*." Pew Center on Global Climate Change, Arlington, VA.

Fischer, Carolyn & Fox, Alan K. (February, 2009). "*Comparing Policies to Combat Emissions Leakage: Border Tax Adjustments versus Rebates*." RFF Discussion Paper No. 09-02, Resources for the Future, Washington, D.C.

GAO. (2008). *Climate Change: Expert Opinion on the Economics of Policy Options to Address Climate Change*, GAO-08-605 (Washington, D.C.: May 9).

GAO. (2008). International Climate Change Programs: *Lessons Learned from the European Union's Emissions Trading Scheme and the Kyoto Protocol's Clean Development Mechanism,* GAO-09-151 (Washington, D.C.: November 18,).

Ho, Mun S., Richard Morgenstern & Jhih-Shyang Shih. (2008). (November) "Impact of Carbon Price Policies on U.S. Industry." RFF Discussion Paper No. 08-37, Resources for the Future, Washington, D.C.

U.S. Environmental Protection Agency. (2009). Office of Atmospheric Programs. EPA Preliminary Analysis of the Waxman-Markey Discussion Draft, The American Clean Energy and Security Act of 2009 in the 111th Congress, Apendix. April 20.

APPENDIX 4

Climate Change Bills Cited in Report

Save Our Climate Act of 2007, H.R. 2069, 110th Cong. (2007).

America's Energy Security Trust Fund Act of 2007, H.R. 3416, 110th Cong. (2007).

Low Carbon Economy Act of 2007, S. 1766, 110TH Cong. (2007).

Lieberman-Warner Climate Security Act of 2008, S. 3036, 1 10TH Cong. (2008).

Investing in Climate Action and Protection Act , H.R. 6186, 110TH Cong. (2008).

Climate Market, Auction, Trust & Trade Emissions Reduction System Act of 2008, H.R. 6316, 110TH Cong. (2008).

Carbon Leakage Prevention Act , H.R. 7146, 110TH Cong., (2008)

America's Energy Security Trust Fund Act of 2009, 111TH Cong. (2009).

Emission Migration Prevention with Long-term Output Yields Act , H.R. 1759, 111TH Cong. (2009).

American Clean Energy and Security Act of 2009 , H.R. 2454, 111TH Cong. (2009).

APPENDIX 5

Glossary

Annex I Countries: Parties to the United Nations Framework Convention on Climate Change (UNFCCC) that are industrialized countries and were members of the Organization for Economic Cooperation and Development (OECD) in 1992 plus countries characterized as economies in transition.

Allowances: In the context of a cap-and-trade system, an emissions allowance is a permit to emit a specific quantity of emissions.

Border allowance requirement: A measure that would require importers to purchase allowances from the federal government prior to importing goods into the United States.

Border tax adjustment: A levy on imported goods proportionate to the imports' embedded carbon content. Generally, the levy on imported goods would be equivalent to the tax applied to domestic goods under a carbon tax system. Three legislative bills-- H.R. 2069, 110TH Cong. (2007), H.R. 3416, 110TH Cong. (2007), and H.R. 1337-- 111TH Cong. (2009), include provisions for a tax on any taxable carbon substance sold by the importer. For the purposes of this report, we are using the term border tax adjustment to refer to this tax described in the bills and to differentiate it from a carbon tax applied to domestic producers.

Business-as-usual: A scenario in which no action is taken to reduce greenhouse gas emissions.

Cap-and-trade: An emissions pricing system in which the government applies an aggregate cap or quota to limit total emissions from regulated sources, which are required to hold allowances to cover their emissions. Allowances are allocated by the government and may be traded. Sources whose allowances exceed their emissions may offer permits for sale, while sources for which emissions exceed allowances will need to buy them.

Carbon dioxide equivalent: The quantity of carbon dioxide emissions that would trap as much heat as a quantity of non-carbon-dioxide gases, such as methane, sulfur hexafluoride, nitrous oxide, or industrial gases.

Carbon leakage: The condition when emissions reductions in one country are replaced by increases in emissions in other countries.

Carbon offsets: Reductions in greenhouse gas emissions from an activity in one place to compensate for emissions released elsewhere.

Carbon tax system: A system that requires regulated sources to pay a charge based on the level of their emissions.

Competitiveness: The ability of a U.S. industry to compete successfully in international markets with foreign competitors.

Cost containment measures: Mechanisms designed to reduce the economic impact of climate change legislation on certain regulated entities and provide them with flexibility in managing compliance costs. Examples include banking and borrowing of allowances, price caps on allowances, free allocation of allowances, and carbon offsets.

Dispute Settlement Body: The WTO's dispute settlement process is administered by the Dispute Settlement Body, composed of representatives from WTO Members, and rules set time limits for each step in the process. .

Downstream regulation: Regulation on sources that emit greenhouse gases.

Embedded carbon content: Carbon emissions associated with the production of a product through the entirety of its supply chain.

Emissions pricing: A market-based mechanism, such as a carbon tax or cap-and-trade system, to encourage reductions in emissions by putting a price on them. In this chapter, "emissions pricing" refers to greenhouse gas emissions pricing, as opposed to pricing systems for other types of emissions.

Energy intensity: The industry's cost of purchased electricity and fuel costs, or energy expenditures, divided by the value of shipments (output) of the industry, as defined in H.R. 2454.

Free allowance allocation: Emission allowances given by the government for free. Under a cap-and-trade program governments can either give allowances to regulated entities for free or they can sell allowances through an auction. Allocating allowances for free represents a transfer of wealth from the government to the entities receiving the allowances, while auctioning allowances enables the government to decide how to use the revenue.

Greenhouse gas intensity: Twenty times the quantity of carbon dioxide equivalent emissions from a sector, divided by the value of shipments (output) for the sector, as defined in H.R. 2454.

Greenhouse gases: Gases such as carbon dioxide, methane, nitrous oxide, and other substances that increase temperatures by trapping heat that would otherwise escape the earth's atmosphere.

Intergovernmental Panel on Climate Change (IPCC): A scientific intergovernmental body set up by the World Meteorological Organization (WMO) and by the United Nations Environment Program (UNEP). The IPCC was established to provide decision makers and others interested in climate change with an objective source of information about climate change.

International reserve allowance: Under a cap-and-trade system, the United States could require importers to acquire emissions allowances corresponding to the level of greenhouse gases emitted during production.

Output-based rebates: Rebates based on actual, recent measures of production that are given to regulated sources in a cap-and-trade system for allowances they have purchased to cover their emissions.

Trade intensity: The ratio of the sum of the value of imports and exports within an industry to the sum of the value of shipments (output) and imports within an industry, as defined in H.R. 2454.

Trade measures: Cost equalization measures at the border that impose a cost or other requirement on energy-intensive imports from countries with weaker climate policies. Depending on the type of domestic carbon mitigation system in place, trade measures can take several forms. For example, trade measures can be proposed as part of a cap-and-trade system (allowance requirement at the border) or a carbon tax system (border tax).

Upstream regulation: Greenhouse gas regulation focused on the sale of fuels that produce greenhouse gases when they are used.

WTO Appellate Body: A body within the WTO's that reviews a WTO panel's legal findings during a dispute resolution process. The Appellate Body and the Appellate Body's report is to be accepted by parties in the dispute unless the Dispute Settlement Body decides by consensus not to adopt the report.

End Notes

[1] A cap-and-trade system is an emissions pricing system in which the government applies an aggregate cap or quota to limit total emissions from regulated sources, which are required to hold allowances to cover their emissions. Allowances are allocated by the government and may be traded. Sources whose allowances exceed their emissions may offer permits for sale, while sources for which emissions exceed allowances will need to buy them.

[2] Under a carbon tax system, regulated sources pay a charge based on the level of their emissions.

[3] You also asked GAO to examine revenue measures under consideration as part of climate change legislation. GAO plans to issue this product later in 2009.

[4] For example, H.R. 2454 (111th Cong.), and H.R. 1759 (111th Cong.) include these criteria. HR 2454 also identifies industries with an energy intensity of 20 percent or greater as vulnerable to adverse competitiveness effects irrespective of trade intensity; however, none of the examples we discuss in our analysis would be identified as vulnerable under these criteria.

[5] A study commissioned by energy-intensive industries identifies 41 manufacturing sub-industries as meeting these vulnerability criteria in at least one year between 2004 and 2006, of which 34 subindustries are within primary metals, nonmetallic minerals, paper products, and chemicals. We did not conduct sensitivity analysis to determine how many additional sub-industries would be identified as vulnerable if the criteria were defined more broadly. According to another expert, the list of potentially vulnerable sub-industries is greater if using the proposed energy-intensive criteria compared with carbon-intensity criteria using a per ton carbon price of $20 or $30. Further, eliminating the trade-intensity criteria would mean that roughly an additional 10 sub-industries would be identified as vulnerable, whereas eliminating the energy-intensity criteria would yield over 150 subindustries that would be identified as vulnerable to competitiveness effects.

In: Carbon Leakage: Trade Issues in Emission Reduction ISBN: 978-1-61728-919-4
Editor: Victoria T. Greenwald

Chapter 4

TESTIMONY OF JOHN J. MCMACKIN, THE ENERGY-INTENSIVE MANUFACTURERS' WORKING GROUP ON GREENHOUSE GAS REGULATION, BEFORE THE HOUSE COMMITTEE ON WAYS AND MEANS, SUBCOMMITTEE ON TRADE, HEARING ON "TRADE APSECTS OF CLIMATE CHANGE LEGISLATION"

John J. McMackin

Mr. Chairman and members of the Subcommittee, it is an honor to be here. The Energy-Intensive Manufacturers' Working Group on Greenhouse Gas Regulation, on whose behalf I appear today, greatly appreciates this opportunity to testify on this critical issue.

I am Jack McMackin, and I am a principal in the law firm of Williams & Jensen, PLLC and a director of Owens-Illinois, Inc. O-I, headquartered in Perrysburg, Ohio and with U.S. facilities in eleven states, is the world's largest manufacturer of glass containers. O-I is a very active and committed member of the Working Group.

Solving the puzzle at the heart of today's hearing is the reason our group was formed. How can we reconcile three things that are seemingly at odds: (1) a unilateral U.S. legislative effort, that (2) addresses a global environmental problem, in light of (3) the reality of global competition? Put differently, our group is all about a solution to "the leakage problem."

I. THE ENERGY-INTENSIVE MANUFACTURERS' WORKING GROUP ON GHG REGULATION- AND "THE LEAKAGE PROBLEM."

The Working Group was formed early last year for a narrow but important purpose: to engage constructively with other stakeholders and Congress to attempt to solve what is often referred to as "the carbon leakage problem" but what is in truth a problem both of the leakage of carbon *and* of jobs. In short, if the U.S. enacts tough global warming regulation but other key manufacturing nations do not, production of energy intensive goods may well shift to the unregulated countries, moving the associated carbon emissions beyond regulation and moving American jobs elsewhere as well.

It is a problem that primarily affects energy-intensive industries that face foreign competition—the two factors that define our members. Our group is composed of companies from the U.S. industries that are widely and correctly seen as most vulnerable to leakage: ferrous metals (iron and steel), non-ferrous metals (aluminum and copper), cement, glass (including fiberglass), ceramics, chemicals and paper. The companies include Alcoa, Corning, Dow, Holcim(US), NewPage Corporation, Nucor, Owens Corning, Owens-Illinois, PPG, Rio Tinto, and U.S. Steel.[1]

I should mention that these are all companies that, of necessity, have already done much to increase their energy efficiency and decrease their emissions. Energy-intensive, trade-exposed industries already have a compelling economic incentive to become energy efficient, which, in turn, leads them to be carbon efficient. Energy costs are a substantial portion of these producers' manufacturing costs. Energy efficiency reduces their cost of energy, which enables them to compete more effectively.

The existence of this incentive is one of the primary reasons that, according to Energy Information Agency Data comparing 1990 emissions to those in 2005, the manufacturing sector as a whole has actually decreased its total emissions, direct and indirect, since 1990, while all other sectors are up, on average, over 30 percent. Similarly, the March 1, 2008 Public Review Draft of EPA's Inventory of Greenhouse Gas Emissions and Sinks 1990-2007 (p. ES-16), shows the industrial sector's total direct and indirect emissions down by 4.2 percent over the period. Moreover, as I will discuss in more detail later in my testimony, our work indicates that based on available data the total emissions of the 40-plus specific industrial sectors or subsectors (by six digit NAICS code) that are most exposed to leakage represent only about 8 percent of total U.S. direct emissions.

The magnitude of potential leakage is a subject of considerable recent study and wide-ranging views. Some of the estimates are truly frightening. In testimony delivered last week before the Energy and Environment Subcommittee of the Energy and Commerce Committee, Richard D. Morgenstern of Resources for the Future gave a longterm figure of 40 percent—assuming a carbon price of only $10: "Over the long term, we estimate that the leakage rate for the few most-vulnerable industries can be as high as 40 percent in the case of a unilateral $10 per ton CO2 price."[2]

There is a broad consensus that the leakage problem must be solved in any responsible global warming legislation. To fail to do so is irrational: it produces economic dislocation and job loss in exchange for no environmental benefit or, even, net environmental harm. The major question as this stage is not whether to address the problem but *how* to address it.

Our group's work is *focused exclusively on one type ofpotential solution*: the grant of free allowances to the most vulnerable manufacturers or the rebating of compliance costs to the most vulnerable manufacturers in some form (including a refundable tax credit). We have not focused, that is to say, on the import ("border equalization") or export ("export rebate") provisions that some have proposed and that the Subcommittee will also be considering. Allow me to make two points that further explain where our proposal fits in the current legislative and policy context.

1. leakage, and Our Proposal to Deal with it, Applies Equally in Either Cap-and-Trade or Carbon Tax Contexts

Leakage results from any form of unilateral U.S. regulation of greenhouse gases that imposes significant costs on energy-intensive, trade-exposed U.S. industries while foreign production is not commensurately burdened. What matters is the climate-policy induced cost differential between U.S. and foreign production of competing goods; the form of the regulation causing the unilateral costs does not matter. Hence, the leakage problem exists whether Congress were to choose a cap and trade structure with its associated carbon "allowance" requirement or a carbon tax.[3] Similarly, the potential remedies to leakage are essentially the same regardless of which form (cap and trade, tax, etc.) mandatory climate policies take.

2. It Is Possible to Divide the Proposed Solutions Into Two Broad Categories (a) "Cost Mitigation" at the Plant Level Which Includes Grants Offree Allowances or Tax Credits, and (b) "Import and/or Export Cost-Equalizing Provisions"

Both of these broad categories seek to lessen the cost that would be imposed by greenhouse gas regulation on U.S. production *relative* to that of unregulated or lesser regulated countries. The first seeks to attack the differential by eliminating or mitigating the cost to U.S producers "at the source," as it were, through free allowance grants, rebates of the cost of allowances, or tax credits to the manufacturer. The second seeks to equalize the costs at the border, either by imposing comparable allowance requirements on imports or by rebating a value-added type tax on exports.

a. Cost mitigation at the source

As indicated above it is the first category, cost mitigation at the source through allowance grants or allowance-value rebates or credits, upon which our efforts are exclusively focused. It is also the principal mechanism adopted by the EU and Australia to deal with the leakage problem that would otherwise be caused by their cap and trade regimes. Moreover, virtually every global warming bill introduced last year had some variation of this form of relief. Its most sophisticated form, and the one that seems to be attracting the broadest support, is an "output based" grant of free allowances (or rebate of allowance value) to energy-intensive, trade-exposed manufacturers.

Output-basing was at the heart of the anti-leakage amendment authored by Senators Brown and Stabenow in Senate consideration of the Boxer substitute to the Lieberman-Warner bill, and is currently the focus of work by Congressmen Inslee and Doyle building on last year's Inslee-Doyle "Carbon Leakage Prevention Act." Last year's bill introduced by Mr. Doggett and his cosponsors, H.R. 6316, the "Climate Matters Act of 2008" contained an allowance grant provision as well as a "border equalization" provision. Moreover, Mr. Doggett, we understand, is currently considering modifying this provision to reflect Inslee-Doyle type output-basing, albeit in the context of a tax rebate or credit.

It is important to note that if a tax code mechanism is used to rebate to leakage- exposed manufacturers some or all of the cost of unilateral greenhouse gas regulation, the provision must be carefully crafted and refundable so that the solution works regardless of a firm's regular or AMT tax status. In any given year, some energy-intensive, trade- exposed industries may not owe income taxes against which a deduction or credit could be applied. Such a tax position could result from any number of factors typical of these companies: eroded revenue or margins from foreign competition and high energy costs, high depreciation from machinery and equipment investments, labor and benefits costs, asbestos-liability payments, and built-up losses and unused credits from any number of sources. If a revenue rebate is not refundable many targeted industries would receive no benefit yet would bear the cost of unilateral regulation—and outcome that does nothing to stop job and carbon leakage.

b. Cost equalization at the border

The second general category has to date largely focused on imports, through "border equalization" provisions that, rather than attempt to mitigate the net cost at the level of the producer instead attempts to impose an equivalent cost on competing products as the border. The provision designed by the International Brotherhood of Electrical Engineers and American Electric Power is a prominent example. The IBEW/AEP proposal seeks to equalize the costs of imports that compete with American energy-intensive goods by imposing a special "international allowance" obligation on such imports. Mr. Doggett's bill from last year contains a parallel provision (as well as its cost-mitigating allowance grant). Likewise, the bill introduced by Mr. Larsen, America's Energy-Security Trust Fund Act, which is structured as a carbon tax, would impose a fee on imports equivalent to the tax.

Other proposals are emerging that attempt to deal with the issue of competitiveness of American exports by structuring greenhouse gas regulation as a charge similar to a value added tax and the rebating that tax on exports.[4] Note that these provisions attempt to "equalize" the cost of American export products with their competitors in non-domestic markets. Unlike the allowance-grant or allowance-value rebates to manufacturers, these export provisions do not attempt to eliminate or mitigate the cost of regulation for energy-intensive goods in the domestic market.

While I reiterate that our Working Group's work has been solely on the first category, the cost mitigating proposals, and we do not as a group take a position on the import/export provisions, I do want to point out that the various approaches are not necessarily incompatible. It is possible to enact both types of provisions in the same legislation, and indeed most of the legislation introduced to date has had both grants of free allowances and border equalization provisions. I will have a few general comments later in my testimony on the relationship of the differing provisions that explain our position that even if the import and/or export provisions are included in greenhouse gas legislation, the cost-mitigating allowance-grant type provisions are still urgently needed.

With respect to the object of our focus, the cost-mitigation proposals, good progress is being made and a convergence is emerging—much of this reflected in the legislation introduced late last year by Congressmen Inslee and Doyle, the "Carbon Leakage Prevention Act"— key provisions of which, as indicated above, are under consideration by Mr. Doggett. Such provisions fit well with bills that focus on tax credit mechanisms as well as more traditional cap-and-trade structures.

II. Good Progress Is Being Made-We Support the Inslee-Doyle Output-Based-Rebate Type Solution, Subject to Further Work on the Eligibility Mechanism and Other Issues. We Urge the Trade Subcommittee and Full Committee to Consider Including Similar Provisions in the Committees' Global Warming Legislation

The legislation Congressmen Inslee and Doyle introduced last year, H.R. 7146, represents the core of a workable solution, and we support its approach. It is not perfect from our point of view and we know it is not final. It should be noted that it certainly does not negate all of the cost that would be imposed by cap and trade legislation on trade-vulnerable, energy-intensive manufacturers. It is appropriately neither a categorical exemption nor a complete elimination of compliance costs. Likewise, it is structured to be transitional relief that keeps American businesses competitive until global agreement can be reached. We are working with the congressmen and other stakeholders to refine it further.

There remain important issues we believe must be addressed. Chief among them is the manner in which last year's bill dealt with selection by EPA of eligible industries. We believe that Congressmen Inslee and Doyle are reworking this section, and we are very hopeful that a new provision will make the process more certain, more objective and more data driven. In any event, I discuss our eligibility concern and a potential solution in more detail later in my testimony (Section II). I would also note that among the other important issues that merit further attention are the fact that the allowances would not be sufficient to cover the full amount of the costs at issue and the amount of discretion to reduce or eliminate the program. I also note that the Dingell-Boucher discussion draft released last year adopted much of the Inslee-Doyle structure as it existed at the time, along with some changes that we think helped advance the thinking of all of us on the leakage problem.

Moreover, I want to stress the "convergence" that we are seeing. In the Senate, those members who have worked most intensely on the issue, such as Senators Brown and Stabenow (as reflected in their amendment in Senate consideration of the Lieberman- Warner bill), as well as many of those in the environmental and academic communities who have studied the issue, USCAP, and others, are not only supporting allowance-grant relief to energy-intensive, trade-exposed industries, they are supporting key structural elements that also undergird the Inslee-Doyle approach. The most important of these is basing allocations on actual output as opposed to historic or grandfathered levels and incorporating an efficiency standard into the allocation formula.

III. Key Features of the Inslee-Doyle Output-Based-Rebate Solution

In essence, the Inslee-Doyle solution, like the Brown-Stabenow solution in the Senate, is a cost-mitigating program that (i) grants free allowances or *rebates allowance value* to energy-intensive trade-exposed industries to compensate them for (ii) a significant portion of the direct allowance and increased electricity costs of a cap-andtrade regime, (iii) that varies the grant based upon a facility's actual, not historic, output, (iv) that rewards a facility for carbon efficiency and punishes it for inefficiency though use of a benchmark or efficiency standard, and (v) that phases out only as international agreements solve the underlying cost disparity.

I will not discuss each of these features in detail, but I do want to note a few of their most important aspects.

A. Output-Based Allotments

The Inslee-Doyle mechanism provides for output-based allotment of allowances. Most of the early anti-leakage, cost-mitigating provisions based their allocation of allowances on a facility's historic emissions. This raised a number or problems, including the following two.

First, historic or grandfathered emissions approaches provide a disincentive to increase production and also discourage new-firm entry—and lost production opportunities in the U.S. may result in production of the same goods elsewhere. An historic-based allocation would not mitigate the cost of additional production. Additional production would be fully exposed to the cost of allowances. Hence, the mechanism would do nothing to help energy-intensive industries to expand production and add jobs. At risk, for instance, would be added production to supply steel, aluminum, copper, glass, ceramics, fiberglass, etc. to what we all hope will be increased production of green products, from wind turbines to solar panels. Similarly, American suppliers would less likely to be providing the cement, plate glass or fiberglass going into new construction of energy-efficient buildings or renovations of older inefficient ones.

Second, some believe historic-based allocations—but not output-based allocations—may in some instances produce an incentive to raise prices but not production. To some commentators this strange phenomenon is a function of "opportunity cost." In some (limited) circumstances, a producer may be able to obtain higher prices, or fail to pass through the cost savings occasioned by free allowances, by in essence saying that if it does not receive from its customers an incremental return on its allowance-grant asset it will reduce production and sell the freed-up allowance. In other words, the existence of this opportunity to sell the allowance changes the seller's supply curve. *In any event, basing the allowance grant on actual output solves this problem—to the extent it exists— by removing the "opportunity" to sell an unused allowance*. A facility only gets an allowance for a product it produces.

Output-basing has another big advantage. It facilitates the use of a benchmark or efficiency standard. A facility's actual production can be included in a formula with an efficiency standard to determine the number of allowances granted.

B. Efficiency Standards

As I described earlier, energy-intensive, trade-exposed industries already have a compelling economic incentive to become energy efficient, which, in turn, leads them to be carbon efficient. That incentive has resulted in remarkable production innovations and efficiency gains. Nonetheless, some policy makers have sought additional assurances that anti-leakage provisions will further incentivize emissions reductions.

Last year's Inslee-Doyle legislation provided this through the use of an effective and practical benchmark: the average energy efficiency of a sector or subsector. This standard has the advantage of being both relatively easy to determine, by definition achievable, and constantly increasing over time. Companies above the average would do relatively better and those below relatively worse, creating an added incentive for each group to improve its efficiency—and thus raising the average. This mechanism inherently rewards operational efficiency and therefore creates a lasting incentive for continuous innovation and technological development.

So, we support the efficiency standard in Inslee-Doyle as introduced last year. We are very concerned, however, about some changes proposed to it. Some would seek to replace the sector-average standard with a "best practices" standard. It would be impossible for companies facing the threat of leakage, or legislators assessing policy options, to know at this juncture whether that which would be deemed by EPA to be the "best practice" in a sector or subsector is economically feasible, or, for that matter, reasonably available.

If it were not, the leakage relief afforded by the allocation grant provision could be illusory. For example, while paper mills use biomass as fuel, many are also coal-fired.

A coal-fired paper plant in Maine, for example, might be forced to close if EPA determined that gas combustion or biomass was the "best practice." And, the jobs lost in all likelihood would not move to a gas-fired plant in the U.S., but, rather, to foreign producers. A best practices regulatory regime is a very different animal than a cap and trade scheme, and attempting to combine the two is very likely a bridge too far. We would strongly oppose it.

C. Direct and "Indirect" Costs

The Inslee-Doyle provision compensates for some, but not all, of the costs that would be imposed by cap and trade legislation. I believe it is important to understand the compromise it represents in this respect.

The costs imposed on U.S. manufacturers by greenhouse-gas legislation will be both those that result directly from their obligation to buy and submit allowances (or under a carbon tax to pay the tax) and "indirectly" from higher prices for electricity, feedstocks, and other production inputs. Moreover, the cost of natural gas, as one example, is likely to increase far more than the cost of allowances associated with its combustion because of the effect of fuel substitution that will drive up the demand for natural gas and because of a shift in the demand curve for natural gas that results precisely from its carbon advantage.

A true cost-*negating* anti-leakage provision would address all indirect as well as direct costs. The Brown-Stabenow amendment in the Senate attempted to take this approach. The Inslee-Doyle cost-*mitigation* approach does not. It would compensate for the increased cost of purchased electricity, but would not compensate for cost increases of feedstocks/inputs, nor would it compensate for the demand and demand-curve caused increases in natural gas. Additionally, Inslee-Doyle imposes an across–the-board 15 percent reduction on its reimbursed costs, direct and indirect—compensating, that is, for only 85 percent of those costs. This was done in part to reduce the grants awarded to a highly efficient producer as a result of the efficiency benchmark.

D. Termination Tied to International Solutions

If the allowance-grant program were to expire on a date certain, or decline on a fixed basis, leakage could re-emerge even after it appeared under control. In fact, because manufacturers need certainty and because they plan their capital allocation far in advance, an expiring anti-leakage provision may well tilt plant location decisions toward foreign locations without regulation. Moreover, a set expiration date would give other countries an incentive to drag their feet in negotiations—to wait us out.

Instead, targeted assistance to energy-intensive industries should be terminated only when the carbon leakage problem is solved through an international agreement. And, it should be phased down only in proportion to progress made in reducing the cost differentials between trading partners in a fashion that demonstrably reduces the disadvantage to domestic producers—not according to an arbitrarily defined timeline. While further refinements are needed, the Inslee-Doyle proposal generally takes this approach.

IV. The Issue of "Qualifying" Industries or Sectors

The Working Group's major issue with the Inslee-Doyle Carbon Leakage Prevention Act as introduced last year concerned its procedures and standards for determining which sectors or subsectors would be eligible to receive allowances. The bill assigned this determination to the EPA subject to a set of criteria that left much room for interpretation and disagreement. In effect, EPA and manufacturers would have been subjected to a series of contested, forecast-rich procedures covering scores of manufacturing sectors and subsectors. These proceedings would be filled with questions of market and product definition as well as competitive impacts. The bill established a very uncertain process—and affected industries need some reasonable level of certainty in making capital expenditure decisions, decisions they must make even now. Similarly, members of Congress from manufacturing states need to know whether their industries will get relief or not.

By contrast, most of the other legislative proposals from last Congress, including the Lieberman-Warner bill, the Boxer substitute, the Brown-Stabenow amendment, and the Dingell-Boucher discussion draft, listed specific industries that would be eligible.

A middle way offering several advantages has emerged. It was suggested by an analysis of the European approach and the work of any number of organizations and scholars—for example, the work of the Peterson Institute and the World Resources Institute in their publication.[5]

Our Working Group has been actively engaged in providing analysis and ideas for this proposal, and it is likewise under consideration by Congressmen Inslee and Doyle. In brief, the provision establishes "presumptive" eligibility through a two-factor test, energy intensity measured by a ratio that sets energy costs over value of shipments and trade exposure measured by the value of imports and exports over the value of shipments plus imports. If a sector or subsector met the presumptive-eligibility standards, it would be eligible for allowances unless the Administrator found that it was not subject to substantial leakage. Any sector or subsector that did not meet the presumptive eligibility tests would be able to establish eligibility through a demonstration of the likelihood of leakage. The actual amount of allowances granted would be decided by the Inslee-Doyle formulas which focus on GHG emissions. The proposed eligibility methodology would make the process of designation of eligible sectors more certain, manageable, principled and data-driven.

V. Some Key Metrics: The FTI Study

Attached to this testimony is a summary of the results of a study by FTI Consulting. We believe and hope it will make an important contribution to analysis of the eligibility issues by all concerned. One of its principal contributions, we think, is to "disaggregate" the very broad categories of industries that had been studied by others and to examine the data at a six-digit North American Industrial Code System level. In addition, it applies objective energy-intensity and trade-intensity criteria to the broad range of American industry, identifying sectors or subsectors that should at least presumptively qualify for relief but that were not on the list most frequently identified. We invite and welcome comment on the study, and we will ask Rob Fisher of FTI to be available for those who wish to work with him.

The study examines the publicly-available trade, energy use, and sales revenue data and implements an energy-intensity threshold of 5 percent and a trade-exposure threshold of 15 percent to determine presumptive eligibility. Both of these standards are consistent with, but somewhat more conservative than, other work to date in the area. For instance, the 5 percent appears to be very near the standard applied by the Peterson/WRI analysis cited above and presents a stricter eligibility threshold than the 4 percent level cited by the recent Pew Congressional Policy Brief, "Addressing Competitiveness in U.S. Climate Change Policy." The study uses the same formula to determine a trade-exposure ratio as does the EU's regulatory scheme, but the study applies a stricter 15 percent trade- exposure compared to the EU's 10 percent.

The FTI study finds 40-plus sectors or subsectors that would qualify for presumptive eligibility, including the list commonly identified as most at risk and represented by our Working Group members. However, the study also identified smaller industries, largely overlooked to date, that meet the criteria and thus would be presumptively qualified. For example, nitrogenous fertilizer with an energy intensity of 14 percent and a trade intensity of 86 percent would qualify, as would wet corn milling, which includes corn sweeteners, atl 1 percent energy intensity and 20 percent trade intensity. The manufacture of refined beet sugar (7 percent energy intensity; 22 percent trade intensity) would qualify as well.

While the energy-intensity and trade-intensity data is relatively straightforward, figuring out the amount of emissions implicated takes considerable extrapolation, so the numbers that follow are approximate. In all, 45 industries are identified as presumptively qualifying (out of the 473 industries included among the NAICS industrial manufacturing codes). These represent about 8 percent of total direct U.S. emissions. When all of the emissions associated with their electricity consumption are included, these industries represent about 10.5 percent of total U.S. emissions. An allowance program that compensated them for the cost of their direct emissions and increased cost of electricity would require about 13 percent of allowances available, for example, under the Lieberman-Warner first year cap of 5,700 million ton CO2e cap in the first year. It should be noted that the 13 percent figure is a rough approximation and that it does not reflect industries that do not qualify for presumptive eligibility but successfully make individuated showings. Moreover, the figure does not include allowances needed to cover production growth.

I should also note that while using data from six-digit NAICS codes to determine whether a sector or subsector would presumptively qualify provides an excellent balance of determinacy, accuracy and administrative ease, it does not work in every instance. Some energy-energy intensive and trade-intensive manufacturing facilities are not classified in six digit codes that meet the presumptive thresholds. For example manufacturers of ceramic substrates for catalytic converters and diesel particulate filters are classified in a NAICS code for auto parts that would not meet the standards. Yet, these manufacturers are energy- and trade-intensive and meet the thresholds on properly individuated data. These circumstances must be accommodated in designing a presumptive-qualification mechanism.

VI. The Relationship of Allowance Grants and Border Cost-Equalization Proposals

As indicated above, while our Working Group's focus has been solely on the Inslee-Doyle type allowance grants that seek to address the leakage problem by mitigating the cost impact of greenhouse gas regulation on energy-intensive and trade- exposed industries, most legislative proposals to date have included border equalization provisions as well. The details of the interface of the two provisions are critical. I wish to make just a few brief and general points about the relationship of the two.

Exports. First, most border equalization provisions suggested so far do not help U.S. manufacturers stay competitive in export markets, and because of the WTO prohibition on export rebates it is difficult to design a border equalization mechanism within a cap and trade structure that will ensure U.S. manufacturers maintain their competitiveness in export markets. Energy-intensive manufacturers are significant exporters. In fact, energy-intensive manufacturing accounts for approximately 14 percent of all U.S. exports. It is unwise to put these exports in jeopardy. This export problem can effectively be addressed through a system of free allowances or rebates without giving rise to a WTO challenge.

Downstream and "Green" Products. Second, border-equalization mechanisms are designed to allow energy-intensive manufactures to pass along the legislation-driven costs to their customers by raising the cost of materials imported into the U.S. by a comparable amount. This creates the troubling possibility that the downstream products could become less competitive as against products produced elsewhere. For instance, because the cost of a bottle is significant part of the cost of a beer or a bottle of wine, Mexican beer and Chilean wine would have a cost advantage over American beer and wines. By way of further example, and assuming the relevant downstream industry is not covered by a border adjustment mechanism, U.S. car assembly plants could be at a cost disadvantage relative to foreign car manufacturing locations that can buy their steel, glass, aluminum and ceramics outside the protective zone of the border equalization provision. This downstream-product phenomenon could be especially harmful to our country's hopes of participating in the manufacture of "green products" such as wind turbines and solar panels. It should be noted that border rebate provisions within a VATtax-like context, which are directed at keeping American products competitive in export markets, likewise do not address this downstream cost problem.

Certainty. Lastly, allowance allocations to energy intensive industries are within our control, are not subject to serious legal challenge, are a feature of cap and trade regimes enacted to date including those in the EU and Australia, and are very unlikely to lead to retaliation or trade wars.

There is a role for WTO-compliant border equalization mechanisms, and perhaps other trade measures, where allowance grants are inadequate or unavailable, and, moreover, such mechanisms should be part of our negotiators' tools. They cannot, however, be the primary mode of relief for the pressing problem of the leakage of carbon and jobs presented by U.S. greenhouse gas regulation.

Mr. Chairman and members of the Subcommittee, thank you very much for this opportunity to appear before you.

ATTACHMENT A

Greenhouse Gas Emissions Legislation Leakage-Exposed Manufacturers Briefing Book

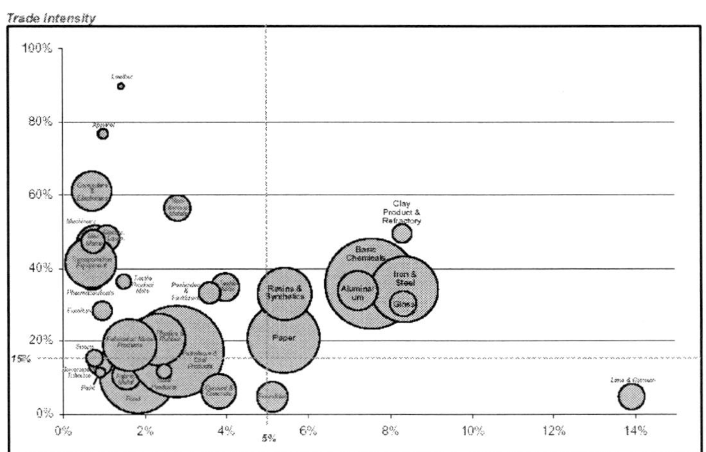

Energy Trade-Intensity = (Imports+Exports) / (Value of Shipments+Imports)
Energy-Intensity = (Energy & Fuel Costs + Generation) / Value of Shipments
Size of Bubble = Amount of energy and fuel consumed as proxy for emissions
Sources: US Census 2006 Annual Survey of Manufacturers, EIA 2002 MECS, US International Trade Commission Tariff & Trade DataWeb

US Manufacturing Leakage Exposure: Aggregate Level

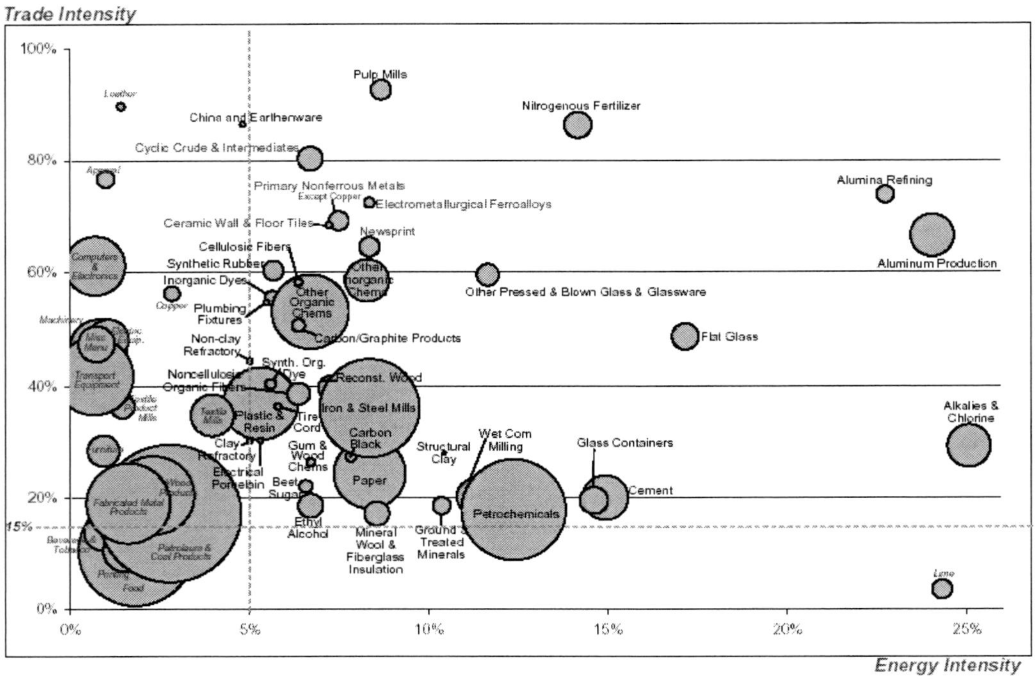

Trade-Intensity = (Imports+Exports) / (Value of Shipments+Imports)
Energy-Intensity = (Energy & Fuel Costs + Generation) / Value of Shipments
Size of Bubble = Amount of energy and fuel consumed as proxy for emissions
Sources: US Census 2006 Annual Survey of Manufacturers, EIA 2002 MECS, US International Trade Commission Tariff & Trade DataWeb

US Manufacturing Leakage Exposure: Detailed Industry Level

LEAKAGE EXPOSURE

To identify the manufacturing industries that are the most highly exposed to this leakage issue, we calculated the energy intensity and trade intensity of each manufacturing sub-industry (at the 6-digit NAICS code).

Methodology for Calculating Energy-Intensity and Trade-Intensity

- We first calculated the energy intensity of each industry. The primary data source we used was the most recent Annual Survey of Manufacturers (ASM) from the US Census, which includes data for 2004-2006. From the ASM we calculated the costs of purchased electricity and of purchased fuel as a percentage of the value of shipments.
- For export and import data, we relied on the US International Trade Commission's Trade Dataweb statistics for 2004-2006, which provides trade data at the 6-level NAICS level.

  - For Imports, we used the US Imports for Consumption category and within that category, we used the CIF Import Value.
  - For Exports, we used the Domestic Exports category. Within that category, we used FAS Value.
- Based on the work in other studies and on the EU's scheme, we have used two thresholds to determine those industries most exposed to leakage[6]:
  1) Energy intensity of **5%**
  2) Trade intensity of **15%**
- The maximum for the last three years of available data (2004-2006) was used to determine eligibility. So, for example, if an industry's energy intensity for the last three years respectively was 4%/5%/4%, it met the eligibility requirement.
- 41 industries met these two criteria.
- In addition, we included NAICS code 311210 (Iron and steel tube and pipe from purchased steel) in order to treat it the same for eligibility purposes as tube and pipe manufactured on an integrated basis.
- We added NAICS code 212210 (Iron ore mining and processing) to capture the beneficiation and other processing for similar reasons – to treat products the same for eligibility purposes whether they are produced at an integrated or non-integrated facility.
- To determine eligibility for the copper industry, we combined the energy and trade data for NAICS codes 331411 (Primary smelting and refining) and 212234 (Copper and nickel mining) to properly capture all copper processing whether the beneficiation of ore occurs at an integrated or non-integrated facility.
- A product that meets the energy intensity and trade intensity criteria should be considered eligible even if the facility that produces it is classified in a non-qualifying NAICS code by virtue of the facility's other products or the facility's ultimate product.

Qualifying Manufacturing Industries

Sectors	NAICS	Sub-Industry	Leakage-Intensity		Value of Shipments
			Energy Intensity	Trade Intensity	
Pulp, Paper & Newsprint Mills	322110	Pulp mills	9%	92%	4
	322121	Paper (except newsprint) mills	8	24	47
	322122	Newsprint mill products	8	64	4
Basic Chemicals	325110	Petrochemicals	12	18	66
	325131	Inorganic dyes and pigments	6	55	4
	325132	Synthetic organic dyes and pigments	6	40	3
	325181	Alkalies and chlorine	25	29	6
	325182	Carbon black	8	27	2
	325188	All other basic inorganic chemicals	8	58	19

(Continued)

Sectors	NAICS	Sub-Industry	Leakage-Intensity		Value of Shipments
			Energy Intensity	Trade Intensity	
	325191	Gum and wood chemicals	7	26	1
	325192	Cyclic crude and intermediates	7	80	9
	325193	Ethyl alcohol	7	18	8
	325199	All other basic organic chemicals	7	53	69
	325211	Plastics material and resins	5	37	79
	325212	Synthetic rubber	6	60	7
	325221	Cellulosic organic fibers	6	58	1
	325222	Non-cellulosic organic fibers	6	38	7
Nitrogenous fertilizer	325311	Nitrogenous fertilizer	14	86	4
Ceramics/Porcelain	327111	Vitreous china plumbing fixtures	6	55	1
	327112	Vitreous china and earthenware articles	5	86	1
	327113	Porcelain electrical supplies	5	30	1
	327122	Ceramic wall and floor tiles	7	69	1
	327123	Other structural clay products	10	28	0.2
	327124	Clay refractory	5	30	1
	327125	Non-clay refractory	5	44	1
Glass Production	327211	Flat glass	17	48	3
	327212	Other pressed and blown glass and glassware; incl. optical fiber	12	59	4
	327213	Glass containers	15	20	4
Cement	327310	Cement	15	20	11
Fiberglass	327993	Mineral wool	9	17	6
Iron & Steel	331111	Iron and steel	8	36	92
	331112	Electrometallurgical ferroalloy products	8	72	1
	331210	Iron and steel pipe and tube from purchased steel			10
	212210	Iron ore mining and beneficiation	18	54	2
Alumina/um	331311	Alumina refining	23	74	1
	331312	Primary aluminum production	24	66	6

(Continued)

Sectors	NAICS	Sub-Industry	Leakage-Intensity		Value of Shipments
			Energy Intensity	Trade Intensity	
Copper	331411	Primary smelting and refining of copper	6	71	10
	212234	Copper and nickel mining and beneficiation			
Other Industries	311221	Wet corn milling	11	20	10
	311313	Beet sugar	7	22	3
	314992	Tire cord and tire fabric mills	6	34	1
	321219	Reconstituted wood products	7	39	8
	327992	Ground or treated minerals and earth	10	19	3
	331419	Primary nonferrous metal (except copper and aluminum)	8	69	5
	335991	Carbon and graphite products	6	50	2

Trade-Intensity = (Imports+Exports) / (Value of Shipments+Imports)

Energy-Intensity = (Energy & Fuel Costs + Generation) / Value of Shipments

Sources: US Census 2006 Annual Survey of Manufacturers, EIA 2002 MECS, US International Trade Commission Tariff & Trade DataWeb EPA -- Inventory of US Greenhouse Gas Emissions and Sinks: 1990-2006

Note: Percentages are rounded to nearest percentage point

Qualifying Industry Descriptions

NAICS	NAICS Name	Description
311221	Wet corn milling	Wet milling corn and other vegetables (except to make ethyl alcohol) to make such products as corn sweeteners (eg, glucose, dextrose, and fructose); corn oil; and starches (except laundry)
311313	Beet sugar	Refined sugar from sugarbeets
314992	Tire cord and tire fabric mills	Cord and fabric of polyester, rayon, cotton, glass, steel, or other materials for use in reinforcing rubber tires, industrial belting, and similar uses
321219	Reconstituted wood products	Reconstituted wood sheets and boards, such as waferboard, oriented strandboard and particleboard
322110	Pulp mills	Pulp manufacturers that do not make paper or paperboard; pulp is made by separating the cellulose fibers from the other impurities in wood or other materials
322121	Paper mills	Paper (except newsprint and uncoated groundwood paper) made from pulp; may also manufacture or purchase pulp
322122	Newsprint mill products	Newsprint and uncoated groundwood paper from pulp
325110	Petrochemicals	Acyclic (aliphatic) hydrocarbons (eg. ethylene, propylene, and butylene) and/or cyclic aromatic hydrocarbons (eg. benzene,

(Continued)

NAICS	NAICS Name	Description
		toluene, styrene, xylene, ethyl benzene, and cumene) made from refined petroleum or liquid hydrocarbon
325131	Inorganic dye and pigments	Black pigments, except carbon black, white pigments, and color pigments.
325132	Synthetic organic dye and pigments	Includes lakes and toners (except electrostatic and photographic)
325181	Alkalies and chlorine	Chlorine, sodium hydroxide (i.e., caustic soda), and other alkalies often using an electrolysis process
325182	Carbon black	Carbon black, bone black, and lamp black
325188	All other basic inorganic chemicals	Basic inorganic chemicals (except industrial gases, inorganic dyes and pigments, alkalies and chlorine, and carbon black)
325191	Gum and wood chemicals	Wood or gum chemicals (eg, naval stores, natural tanning materials, charcoal briquettes, and charcoal, except activated) or Distillation of wood or gum into products (eg, tall oil and wood distillates)
325192	Cyclic crude and intermediates	Cyclic crudes or, cyclic intermediates (i.e., hydrocarbons, except aromatic petrochemicals) from refined petroleum or natural gas or the distillation of coal tars
325193	Ethyl alcohol	Nonpotable ethyl alcohol
325199	All other basic organic chemicals	Basic organic chemical products (except aromatic petroche-micals, industrial gases, synthetic organic dyes and pigments, gum and wood chemicals, cyclic crudes and intermediates, and ethyl alcohol)
325211	Plastics material and resins	Resins, plastics materials, and nonvulcanizable thermoplastic elastomers and mixing and blending resins on a custom basis and/or noncustomized synthetic resins
325212	Synthetic rubber	Synethetic rubber such as Styrene-Butadiene-Rubber (SBR), butyl, polychloroprene, and stereo polyisoprene elastomers
325221	Cellulosic organic fibers	Cellulosic (i.e., rayon and acetate) fibers and filaments in the form of monofilament, filament yarn, staple, or tow
325222	Non-cellulosic organic fibers	Noncellulosic (i.e., nylon, polyolefin, and polyester) fibers and filaments in the form of monofilament, filament yarn, staple, or tow
325311	Nitrogenous fertilizer	Production of fertilizer through inorganic (Synthetic ammonia, nitric acid, urea, and ammonium compounds) or organic sources
327111	Vitreous china plumbing fixtures	Vitreous china plumbing fixtures and china and earthenware bathroom accessories, such as faucet handles, towel bars, and soap dishes
327112	Vitreous china and earthenware articles	Table and kitchen articles, art and ornamental items, and similar vitreous china, fine earthenware, stoneware, coarse earthenware, and pottery products
327113	Porcelain electrical supplies	Porcelain electrical insulators, molded porcelain parts for electrical devices, ferrite or ceramic magnets, and electronic and electrical supplies from nonmetallic minerals, such as clay and ceramic materials
327122	Ceramic wall and floor tiles	Includes mosaic and quarry tiles
327123	Other structural clay products	Clay sewer pipe, drain tile, flue lining tile, architectural terra-cotta, and other structural clay products

(Continued)

NAICS	NAICS Name	Description
327124	Clay refractory	Clay refractory, mortar, brick, block, tile, and fabricated clay refractories, such as melting pots. A refractory is a material that will retain its shape and chemical identity when subjected to high temperatures and is used in applications that require extreme resistance to heat, such as furnace linings.
327125	Non-clay refractory	Nonclay refractory, mortar, brick, block, tile, and fabricated nonclay refractories such as graphite, magnesite, silica, or alumina crucibles.
327211	Flat glass	Flat glass made by melting silica sand or cullet (includes integrated facilities that also produce laminated glass)
327212	Other pressed and blown glass and glassware	Glass made by melting silica sand or cullet and products made by pressing, blowing, or shaping glass or glassware (except glass packaging containers); Also includes fiber optics
327213	Glass containers	Glass containers for commercial packing and bottling, and for home canning, including bottles and jars
327310	Cement	Portland, natural, masonry, pozzalanic, and other hydraulic cements; manufacturers may calcine earths or mine, quarry, manufacture, or purchase lime
327992	Ground or trea-ted minerals and earth	Calcining, dead burning, or otherwise processing beyond beneficiation, clays, ceramic and refractory minerals, barite, and miscellaneous nonmetallic minerals
327993	Mineral wool and fiberglass insulation	Mineral wool and mineral wool (i.e., fiberglass) insulation products made of such siliceous materials as rock, slag, and glass or combinations thereof
331111	Iron and steel mills	Steel production, direct reduction of iron ore, manufacture of pig iron, conversion of pig iron into steel; includes both BOF and EAF; includes integrated facilities that also manufacture shapes (e.g., bar, plate, rod, sheet, strip, wire) or form tube and pipe
331112	Electrometallurg ical ferroa-lloys	Ferroalloys add critical elements, such as silicon and manganese for carbon steel and chromium, vanadium, tungsten, titanium, and molybdenum for low- and high-alloy metals
331210	Iron and steel pipe and tube from purchased steel	Welded, riveted, or seamless pipe and tube from purchased iron or steel
212210	Iron ore mining and processing	Mine site development, mining, and/or beneficiation (i.e., preparation) of iron ores and manganiferous ores valued chiefly for their iron content and/or (2) sinter iron ore production (except iron ore produced in iron and steel mills) and other iron ore agglomerates
331311	Alumina refining	Alumina (i.e., aluminum oxide) refining generally from bauxite
331312	Primary aluminum production	Aluminum production from alumina; includes integrated facilities that also roll, draw, extrude, or cast the aluminum into primary forms (e.g., bar, billet, ingot, plate, rod, sheet, and strip)
331411	Primary smelting and refining of copper	Smelting of copper ore and/or the primary refining of copper by electrolytic methods or other processes to makeprimary copper and copper-based alloys, such as brass and bronze, from ore or concentrates

(Continued)

NAICS	NAICS Name	Description
212234	Copper and nickel ore mining and beneficiation	Mine site development, mining, and/or beneficiation (i.e., preparation) of copper and/or nickel ores, and recovery of copper concentrates by the precipitation, leaching, or electrowinning of copper ore
331419	Primary nonferrous metals (except copper and aluminum)	Primary production of nonferrous metals by smelting ore and/or by electrolytic methods or other processes; includes lead, gold, silver, titanium, zinc and magnesium
335991	Carbon and graphite products	Carbon, graphite, and metal-graphite products including fibers, brushes and brush stock, and electrodes for thermal and electrolytic uses

METHODOLOGY FOR ESTIMATING EMISSIONS

After identifying the manufacturing industries exposed to carbon leakage, we then estimated the emissions for these qualifying industries to determine the scope of required allowances to address the issue. We estimated the 2007 emissions based on available EPA and EIA data. Since sufficient emissions data are not available at a sub-sector level, we needed to make certain assumptions based primarily on electricity and fuel use to estimate emissions for qualifying industries.

Emissions (EPA's Inventory of US Greenhouse Gas Emissions and Sinks: 1990-2007)[7]
(Million metric tons CO_2 Equ.)

	Fuel Combustion	Electricity	Industrial Processes	Other Emissions	Total
Total	3,350	2,397	328	992	**7,068**
Industrial	857	708			1,565
Manufacturing	802	614			1,416
Representative Industries	**391**	**222**	**158**		**771**
% of Total Emissions	5.5%	3.1%	2.2%		10.9%
% of 2012 Allowances	6.8%	3.8%	2.7%		13.4%
(Est. 5,775)					

- We started with the EPA 2007 emissions for the Industrial sector, broken out between combustion and electricity.
- Using the EIA's Annual Energy Outlook 2008, we calculated the percentage of Industrial emissions that Manufacturing represented for electricity (87%) and for consumption (94%).[8]
- The next step was to allocate the overall Manufacturing emissions to each sub-industry of the Manufacturing sector to determine how many emissions our qualifying industries represent. To allocate electricity emissions, we applied the percentage of purchased and generated kWhs of electricity (from the ASM) for each

sub-industry to the overall electricity emissions figure (612 million) for Manufacturing.

- To estimate the consumption emissions, we started with the manufacturing consumption figure derived from the EPA (806 million). We used the EIA's 2006 paper[9] to determine the initial allocation percentages. We first added up the emissions for all the fuel sources and calculated the percentage of the total for each industry. Over 90% of the consumption emissions are allocated to an industry. We allocated to sub-industries where necessary based on the emissions data given and then based on the fuel spend. For the remaining 8.6%, we allocated to the other industries based on fuel spend.
- We assigned/allocated the process emissions from the EPA to the individual industries.
- Definitions of emissions categories:
    - Fuel Combustion: Fossil fuel combustion to generate heat, steam or electricity to power industrial processes
    - Electricity: Indirect emissions from purchased electricity
    - Industrial Processes: Byproduct or fugitive emissions of greenhouse gases from industrial processes not directly related to energy activities such as fossil fuel combustion

Emissions for Qualifying Manufacturing Industries (million metric tons CO_2 Equ.)

Manufacturing Industry	NAICS	Sub-Industry	Fuel Combustion	Electricity	Industrial Processes	TOTAL
Food	311221	Wet corn milling	14	6		19
	311313	Beet sugar	3	1		4
Textiles	314992	Tire cord and tire fabric mills	0	1		1
Wood Products	321219	Reconstituted wood products	2	4		6
Paper	322110	Pulp mills	2	1		3
	322121	Paper (except newsprint) mills	28	20		48
	322122	Newsprint mill products	2	2		3
Chemicals	325110	Petrochemicals	27	10	4	41
	325131	Inorganic dyes and pigments	1	1		3
	325132	Synthetic organic dyes and pigments	1	1		2
	325181	Alkalies and chlorine	11	14	2	26
	325182	Carbon black	5	1		5
	325188	All other basic inorganic chemicals	7	15		22
	325191	Gum and wood chemicals	1	0		1
	325192	Cyclic crude and intermediates	3	3		6
	325193	Ethyl alcohol	6	3		9
	325199	All other basic organic chemicals	52	24	6	82
	325211	Plastics material and resins	46	21		67

(Continued)

Manufacturing Industry	NAICS	Sub-Industry	Fuel Combustion	Electricity	Industrial Processes	TOTAL
	325212	Synthetic rubber	2	1		3
	325221	Cellulosic organic fibers	1	1		1
	325222	Non-cellulosic organic fibers	3	4		7
	325311	Nitrogenous fertilizer	9	3	36	48
Non-Metallic Mineral Products	327111	Vitreous china plumbing fixtures	0.4	0.1		0
	327112	Vitreous china and earthenware articles	0.3	0.1		0.4
	327113	Porcelain electrical supplies	0.2	0.3		0.5
	327122	Ceramic wall and floor tiles	1	0		1
	327123	Other structural clay products	0.2	0.0		0.2
	327124	Clay refractory	0.3	0.2		0.5
	327125	Non-clay refractory	0.4	0.2		1
	327211	Flat glass	3	1	1	4
	327212	Other pressed and blown glass and glassware	2	2	0.4	4
	327213	Glass containers	3	2	1	6
	327310	Cement	29	8	45	82
	327992	Ground or treated minerals and earth	2	1		3
	327993	Mineral wool	2	3		5
Primary Metals	331111	Iron and steel	111	36	54	201
	331112	Electrometallurgical ferroalloy products	1	1		2
	331210	Iron and steel pipe and tube from purchased steel	1	1		2
	212210	Iron ore mining and beneficiation	3	4		8
	331311	Alumina refining	2	0.3	1	4
	331312	Primary aluminum production	4	17	7	28
	311411	Primary smelting and refining of copper	1	0.4		1
	212234	Copper and nickel mining and beneficiation	1	3		4
	331419	Primary nonferrous metal (except copper and aluminum)	1	3	4	7
Electrical Equipment	335991	Carbon and graphite products	0.5	1		1
			391	222	158	771

Fuel Combustion: Fossil fuel combustion to generate heat, steam or electricity to power industrial processes

Electricity: Indirect emissions from purchased electricity

Industrial Processes: Byproduct or fugitive emissions of greenhouse gases from industrial processes not directly related to energy activities such as fossil fuel combustion

Frequently Asked Questions

- **Why use energy intensity rather than emissions?**
  For purposes of identifying qualifying industries, as opposed to calculating the number of allowances to be awarded, energy spending arguably is a better metric than emissions to determine the financial impact to manufacturing industries of greenhouse gas legislation. For instance, firms may be impacted by the increases in the cost of natural gas that will be inversely related to natural gas's relative carbon intensity advantage. Any emissions metric also will depend on an assumed allowance price, which is difficult to estimate. In addition, emissions data is not readily available at a detailed industry level. There are 473 Manufacturing sub-industries (6-digit NAICS). No known resource provides emissions data, or even detailed fuel usage data from which emissions could be derived, at a level of detail anywhere close to that. In any event, energy spending serves as a reasonable proxy for emissions since combustion emissions are a function of the energy used, varying only by the mix of fuels.
- **Why use the Census's ASM rather than EIA's MECS?**
  The most recent data for the MECS Survey is from 2002 rather than from 2006 for the ASM. Additionally, the ASM provides data on a much more granular NAICS level than the MECS Survey does. For the 473 NAICS sub-industries (6-digit), MECS provides energy data for only 39 at the 5- or 6-digit level. ASM provides 2005 or 2006 energy data for 472 of the 473 industries (Petrochemicals is the only one with no data) at 5- or 6-digit level. We have validated our results against MECS and in the case of Petrochemicals, we used the MECS data.
- **Why is Imports in both the numerator and the denominator of the trade intensity formula?**
  The size of any US market in which a manufacturer competes equals Domestic Production + Imports, where Exports are a subset of Domestic Production. Using Imports in the denominator of the trade intensity formula keeps the ratio from exceeding 100%. Note: We have used the same ration that is used by the EU.
- **How is the variability from year to year in energy intensity and trade intensity addressed?**
  We used the maximum of the three years (2004-2006) for energy intensity and trade intensity, which minimizes situations where an industry may qualify one year and not another. However it should be noted that these metrics have very little variability from year to year. The average annual change in energy intensity is only 0.2%. The average annual change in trade intensity is only 3%.

End Notes

[1] While this written testimony generally represents the position of the Working Group, not all statements are necessarily endorsed by every member. I do not represent members of the group other than Owens- Illinois, and while my responses to any questions during the hearing will attempt not to stray from the group's views, those responses will be my own and not necessarily the group's.

[2] Testimony of Richard D. Morgenstern, Senior Fellow, Resources for the Future, Prepared for the Committee on Energy and Commerce, U.S. House of Representatives, Hearings on Competitiveness and Climate Policy:

Avoiding Leakage of Jobs and Emissions, March 18, 2009, at 5. On page 7-8 of his testimony Mr. Morgenstern cautions that the modeling done by him and his collaborators may somewhat overstate leakage because some of the trade partners it includes, such as the EU, have themselves adopted carbon regulation. However, most of these, it should be noted, such as the EU and Australia, have themselves adopted anti-leakage allowance-grant programs for their energy-intensive and trade exposed industries.

[3] For a discussion of the leakage problem in a carbon tax context see generally, Metcalf and Weisbach, "The Design of a Carbon Tax," Working Paper 09-05, January 2009, AEI Center for Regulatory and Market Studies (2009).

[4] For a general discussion of the various forms of import and export provisions see, Fischer and Fox, "Comparing Policies to Combat Emissions Leakage: Border Tax Adjustments versus Rebates," Discussion Paper, Resources for the Future (February 2009).

[5] Houser, Trevor, Rob Bradley, Britt Childs, Jacob Werksman, and Robert Heilmayr, *Levelling the Carbon Playing Field: International Competition and US Climate Policy Design,* May 2008.

[6] Ratios were rounded to nearest whole percentage point. For example, a calculated energy intensity of 4.51% or higher was rounded up to 5.0% and thus met the 5% threshold.

[7] The Total and Industrial rows are from the EPA, remaining data is derived

[8] The Industrial sector is Manufacturing plus Agriculture, Forestry, Fishing, Hunting, Mining and Construction

[9] Energy-Related Carbon Dioxide Emissions in US Manufacturing (November 2006; Mark Schipper)

In: Carbon Leakage: Trade Issues in Emission Reduction
Editor: Victoria T. Greenwald
ISBN: 978-1-61728-919-4

Chapter 5

TESTIMONY OF LEO W. GERARD, INTERNATIONAL PRESIDENT, UNITED STEELWORKERS, BEFORE THE HOUSE COMMITTEE ON WAYS AND MEANS, SUBCOMMITTEE ON TRADE, HEARING ON "TRADE ASPECTS OF CLIMATE CHANGE LEGISLATION"

Leo W. Gerard

Good afternoon. On behalf of the 850,000 active members of the United Steelworkers (USW), I would like to thank Chairman Levin for holding this hearing on the challenges to the competitiveness of domestic manufacturers and workers posed by the adoption of comprehensive climate change legislation. I am Leo Gerard, the International President of the USW. As you know, the members of the United Steelworkers produce more than just steel. They supply almost every sector of the economy, including the North American auto industry, and produce a wide array of products, including paper, glass, ceramics, cement, chemicals, aluminum, tires and rubber. Our members produce these energy-intensive products in facilities that are as efficient as any in the world. They are ready to answer the call to produce the next generation of clean energy products and parts, and reassert America's leadership on the cutting edge of new technology. But they can only answer that call if their jobs are not unnecessarily squandered to the law of unintended, but not unforeseen, consequences. Amid this economic collapse, this country cannot afford to lose any more jobs.

For decades, the USW has been a leader in the labor movement on the environment. In 1990, we published "Our Children's World" stating our union's environmental policy and the need to address climate change, and in 2006 we reaffirmed our union's commitment to environmental responsibility through the publication of "Securing Our Children's World."[1] We were one of the first industrial unions to support comprehensive climate change legislation, with our support for the Bingaman-Specter bill. That bill proceeded from recommendations made by the National Commission on Energy Policy, on which I serve as a commissioner. USW is also a founding member of the Blue-Green Alliance, which brings together unions and environmental groups to plan a new way forward for America through the promotion of policy solutions that spur growth and investment in green technologies and products produced here in America.

The Steelworkers are as convinced today as we were in 1990 that climate change is the most important environmental issue of our lifetime. It is the challenge of our time to transform the way this nation operates in order to bring this problem under control before it is too late. Still, in undertaking the enormous and critical task of crafting comprehensive climate change legislation, Congress must ensure that the desired emissions reductions are achieved in a structured, responsible way. The legislation must not only strive to reduce emissions to the level that the best science believes is necessary, but it must do so in a way that minimizes costs to businesses and consumers as much as possible. In doing so, attention must be paid to the need to provide incentives to build the next generation of clean energy products here in America, and the need to ensure that domestic exporters are not unfairly disadvantaged in the global marketplace. It must take into account that, for some products like steel and cement, some emissions are an unavoidable part of the manufacturing process, and that currently neither science nor technology exists to mitigate them. And it must ensure, as much as possible, that the jobs that exist here today in energy-intensive manufacturing are not lost, nor the production of those products offshored unnecessarily by neglecting the very real and potentially disastrous problem of carbon leakage. If leakage is not addressed in the development of a climate change regime, any policy runs a significant risk of not only costing American jobs but actually exacerbating, instead of mitigating, the problem of global warming.

Carbon Leakage

The phenomenon by which emissions reductions in one country lead to increased emissions in another is known as carbon leakage. The reason this happens is that if one country puts a price on carbon emissions, that additional cost provides an incentive to the company to move its production and, therefore, its emissions, to a country where that additional cost does not exist. All policy proposals to address climate change, including cap-and-trade, arise from the idea that if a price is put on carbon, it will provide an incentive to emit less carbon. This theory is sound, as long as the cost cannot simply be evaded by companies moving production overseas or by downstream producers and consumers avoiding the cost by purchasing imported materials from nations that do not share the U.S.'s commitment to climate change abatement.

This threat of leakage is particularly acute among manufacturers of energy-intensive primary products like the ones made by members of the Steelworkers. In commodity-based industries like steel, glass, chemicals, rubber, and paper, even small differences in production costs can devastate an industry if they are not managed effectively. Finding a way to mitigate the competitive disadvantage that will be placed on these industries is not only an imperative, if we are to continue the recovery from the current recession, but it is an imperative if we are to actually achieve the goal of stopping climate change.

Greenhouse gas emissions and the resulting climate change are a global problem, and it makes no difference whether the emissions occur here in the U.S. or abroad. In fact, the shifting of these emissions to countries that do not share our commitment to addressing the problem of climate change is almost certain to make the overall problem worse. The reason for this is quite simple: American industry and American workers are among the best in the world, and they produce energy-intensive goods with some of the lowest emissions in the world. The same cannot be said of many of our competitors. The Alliance for American Manufacturing, a unique labor-management joint venture between the Steelworkers and several of our major employers, released a report yesterday on the pollution levels in the Chinese steel industry, and the findings are quite stark.[2] For example, while the American steel industry has become 25% less energy intensive over the past 20 years, the Chinese steel industry now emits as much carbon as the rest of the global steel industry combined. The production of a ton of steel in China generates more than three times the carbon emissions of a ton of steel produced in the United States. This is largely because the domestic industry is increasingly state-of-the-art and efficient, while the Chinese steel industry has a heavier reliance on older, dirtier production methods and uses higher-sulfur coal to power those processes. The Chinese government looks the other way while this goes on, and is lax in enforcing the few environmental laws and regulations it does ostensibly have in place.

Any climate change policy that does not seek to prevent the unnecessary offshoring of production from state-of-the-art American industries to less efficient, more carbon-intensive industries overseas will both cost American jobs and, perversely, will actually make the problem of global climate change worse.

Options for Combating Leakage

The USW is pleased that a growing consensus is forming around the idea that something must be done to address the leakage problem in formulating climate change policy. The question that follows is exactly what that something should be. A variety of solutions have been proposed, many of which fall into the broad categories of allocation schemes and trade mechanisms.

Allocations

Because leakage is caused by the fact that the domestic industry will be bearing increased costs of production due to the requirement to pay an imposed cost of carbon, many proposed solutions center around the concept of mitigating those costs. These ideas are structured as allocations of allowances to industries that are at risk of leakage, which means energy-intensive and trade-exposed industries. The European cap-and-trade program relies exclusively on allocations to combat leakage.

Previous domestic efforts, such as the 2008 Lieberman-Warner bill, have included provisions that reserve a certain percentage of the total universe of allowances to be distributed to energy-intensive industries free of charge. This structure is less than ideal because the allocation of no-strings allowances provide little incentive to companies to avoid offshoring. The potential for a company to take its free allowances, sell them on the allowance market, and use the windfall profits to build factories in India, Mexico, Brazil or China is a serious concern. In addition, even those companies that use the allocations as intended still face a long-term leakage threat. Most allocation proposals decrease the percentage of the cap reserved for allocations over time, which would allow foreign competitors to wait out their domestic counterparts until the supply of free allowances runs out. Even those proposals that maintain a consistent percentage of the cap for allocations face the same problem, as the cap will get smaller and smaller, as will the total number of available allowances the consistent percentage represents.

While allocations are critical for the survival of energy-intensive manufacturers, they must be structured to provide an incentive to maintain or increase domestic production, and must eliminate the potential for windfall profits, particularly profits which can be used to facilitate offshoring.

Trade Mechanisms

Where allocation schemes seek to even out the cost differential between domestic and international products by reducing the effective cost to domestic producers, trade mechanisms do the opposite. An effective trade mechanism would eliminate the cost differential by requiring that any import that enters our market face the same cost as domestic counterparts for those emissions not covered by an allocation scheme.

The most prominent of these proposals is the international reserve allowance program in the Lieberman-Warner bill. Between the introduction of the bill and the version improved by Senator Boxer, the international reserve allowance program was refined and improved a great deal, but more work needs to be done before it can fully address leakage concerns. A workable trade mechanism must give consideration to downstream products and exports. It must require that all products consumed in the U.S. demonstrate the same commitment to combating climate change, no matter where they are produced. And it must be put in place as quickly as possible, to limit the amount of time that domestic producers face cost disadvantages because of the requirements of the domestic program. If it is not possible to begin both programs at the same time, then steps must be taken to prevent unnecessary harm to domestic industries until such time as the trade mechanism can be activated.

Access to our consumer market is the most powerful incentive the U.S. has to encourage other nations to commit to reduce climate change. It must be used in a strong and effective manner.

Hybrid Approach

The shortcomings of both the allocation approach and the trade approach are similar. Namely, this is a global economy that faces a global crisis, and there are limits to what any one country, even the United States, can do alone. The U.S. should, therefore, attempt to forge a global solution to the issue of how to deal with energy intensive manufacturers. This should take the form of global sectoral agreements within the larger global climate treaty being negotiated by the U.N. Framework Convention on Climate Change. Only by setting up a system where all products must bear a carbon cost commensurate with its carbon emissions, no matter where they are produced, can the playing field ever be truly leveled and allow us to confront this global problem.

With that as the long-term goal, the short-term goal should be to craft a hybrid approach of allocations and trade measures that increases the potential that such agreements can be reached, while still addressing the leakage and competitiveness questions and ensuring that industry has sufficient incentive and confidence to maintain domestic production here, while continuing to improve its operations, until such agreements can be reached.

In this hybrid approach, allocations could be awarded to energy intensive manufacturers commensurate with their output and their carbon emissions. If allocations diminish over time or are insufficient to eliminate the leakage problem, they can be combined with appropriate border adjustments to equalize costs for domestic and foreign goods consumed in the United States based on their associated emissions. A phased-in, hybrid approach could provide the space for both the negotiation of an international agreement – which should start upon passage of the legislation – and providing sufficient notice to the rest of the world of the eventual imposition of a meaningful trade mechanism, while preventing domestic producers from facing unnecessary competitive pressures during that time. In addition, the hybrid approach can be designed to address the problems of downstream products and exports by ensuring that costs to inputs are minimized, and thus downstream products do not see an additional cost disadvantage. Similarly, if exported goods do not face a disadvantageous cost differential abroad, their competitiveness in global markets should not be harmed.

After the negotiation period is over, a variable border adjustment will be imposed on imports. This adjustment will be imposed on imports that enjoy a cost advantage over domestic products because of lack of action on climate change. It will be based on the carbon intensity of these products and the net cost borne by domestic manufacturers of those same products.

It is a simple concept. The right to sell goods to consumers in our market brings with it the responsibility to confront the costs associated with addressing climate change.

If the output-based rebates are working as intended and meeting the competitiveness needs of energy-intensive manufacturers, the border tax adjustment will lay dormant. Similarly, if sectoral agreements are forged and work as intended, this will be a tax that no one has to pay. That is the goal, and the border tax adjustment is envisioned to be a last resort, put into use only if and when the allocations are insufficient, or the sectoral agreement is not enforced.

An Alternative Approach

Hybrid approaches, allocation schemes, and trade mechanisms that could face WTO challenges are all quite complicated ways to address the questions of leakage and competitiveness. The questions themselves largely stem from the fact that the architecture of a cap-and-trade system is focused on the production of goods, but the global economy is focused on the consumption of goods. An alternative approach for energy-intensive manufacturers would be to create a separate emissions regime for these industries in which the inefficient allowance-based system is replaced with a simpler and more effective system in which emissions fees are assessed on all carbon-intensive goods consumed in the U.S. if their associated carbon emissions exceed a determined industry standard.

The potential benefit of such a system would be that the leakage problem would be effectively eliminated, because the focus would be shifted to ensuring that all products consumed in the U.S., regardless of where they are made, demonstrate the U.S.'s commitment to addressing climate change. Domestic manufacturers would face incentives to reduce emissions in order to bring emissions under the standard and avoid the tax. At the same time, they would not face unnecessary competitiveness concerns because equivalent costs can be assessed at the border on imports and rebated on exports, in much the same way as a value-added tax. In addition, the transparency of these fees would help industry attract the necessary capital to make improvements, because future costs could be more easily determined using an established fee rate than in attempting to divine the price of a volatile market in carbon allowances.

Conclusion

Addressing the potentially catastrophic issues posed by climate change is the challenge of our generation, and meeting that challenge will require the mobilization of everyone in the world behind a common purpose. America can and must lead this effort, not only by taking a bold stand to limit greenhouse gas emissions, but by harnessing this nation's greatest resource, the ingenuity and creativity of the American people. We must make a national commitment to rebuild America clean and green with products built here, to develop new forms of clean, renewable energy and provide incentives to further their deployment. We must bring our power grid and energy infrastructure into the 21st century and train the American workforce to use these new technologies. We must create a revolution in our transportation sector, rebuilding the American auto industry to produce the best and cleanest vehicles in the world, and connect America's cities and neighborhoods with world class transit systems. And, of course, we must limit greenhouse gas emissions consistent with what the best science tells us.

In creating a program to achieve these emissions reductions, we must make the development of manufacturing a centerpiece of that program. The products made by our members and millions of other hard-working Americans are quite literally the building blocks of all these new technologies. If the U.S. is to build windmills, we will need steel and aluminum. If we are to build solar panels, we will need glass. And if we are to build the next generation of industrial scrubbers to filter out these emissions, the ceramics industry cannot be ignored.

When the world transitioned to an industrial economy, America led the way by developing and producing the best products in the world. Now, as the world transitions again to a green economy, the time has come for America to lead again. This change will not come easily, and it is a heavy load to bear. But I am here to tell you today that American workers are ready and willing to help bear that burden and help lead America into a new, green future.

Thank you again, Mr. Chairman, for holding this hearing. The United Steelworkers and I look forward to working with you and the committee to renovate our economy to meet these challenges.

[1] Available on USW's website through the following links; http://legacy.usw.org/uswa/program/content/1592.php and http://legacy.usw. org/usw/ program/ content/Environment-SOCW.php.

[2] Available on Alliance for American Manufacturing's website through the link http://www.americanmanufacturing.org/assessment-of-china.

In: Carbon Leakage: Trade Issues in Emission Reduction ISBN: 978-1-61728-919-4
Editor: Victoria T. Greenwald

Chapter 6

TESTIMONY OF DAVID HAMILTON, DIRECTOR OF GLOBAL WARMING AND ENERGY PROGRAMS, SIERRA CLUB, BEFORE THE SUBCOMMITTEE ON WAYS AND MEANS, HEARING ON "THE POTENTIAL EFFECTS OF CLIMATE CHANGE LEGISLATION ON THE COMPETITIVENESS OF DOMESTIC ENERGY-INTENSIVE INDUSTRIES"

David Hamilton

Thank you for the opportunity to address the Subcommittee on the importance of providing a fair and effective system to address the potential effects of a carbon control regime on U.S. energy intensive industries and their international competitiveness.

My name is David Hamilton and I am the Director of Global Warming and Energy Programs for the Sierra Club. The Sierra Club is the nation's oldest, largest, and most influential environmental organization with 1.3 million members and supporters. We have chapters in 50 states and tens of thousands of volunteers across the country that put their personal time into protecting and improving the environment of their communities, states and the nation as a whole.

We are at a unique and very difficult point in history. During a time of economic emergency, we must address our emissions of pollution that is rapidly warming the planet and disabling many of the biological systems that have enabled life on Earth. This challenge also represents a unique opportunity to foster sustainable growth as we transition into a green energy economy, creating millions of jobs in the process.

STEMMING GLOBAL WARMING IS SIERRA CLUB'S TOP PRIORITY

After focusing most of its efforts for more than a century primarily on protecting wild and special places, the Sierra Club chose global warming in 2005 as its top priority issue. That decision is continually reinforced by somber news from the scientific community. Nearly every day, more research emerges on the progression of warming and other aspects of climate disruption that the Intergovernmental Panel on Climate Change (IPCC) has identified as very likely resulting from the build-up of carbon dioxide and other gases emitted into the atmosphere through human activity.

We are in a race against time to reduce these emissions and slow their effects on the planet. As NASA scientist James Hansen and others have articulated, the continued warming of the planet will bring us to tipping points such as the thawing of Arctic tundra that lead to more and uncontrollable emissions of carbon dioxide – thus leading to more warming. If we lose this race to bring carbon- based warming under control, its effects will be out of our hands and all life on Earth will be faced with a severely altered home that will challenge nearly every aspect of society.

A GREEN ECONOMY: COMBATING GLOBAL WARMING AND CREATING JOBS

The Sierra Club has regarded the economic health of American wage-earners as a pre-condition for the nation's environmental health for many years. We have worked shoulder to shoulder with our partners the United Steelworkers and other unions to fight for real protection for both workers and the environment in international trade agreements going back to the development of North American Free Trade Agreement (NAFTA). We formed the Blue-Green Alliance with the Steelworkers that now includes the Communications Workers of America, the Service Employees International Union, and the Laborers International Union of North America, as well as Natural Resources Defense Council. In a steady and inexorable process, blue and green together increasingly recognize that quality of life for Americans is determined by both economic and environmental security.

I want to thank you Mr. Chairman, for your leadership in creating a new vision for our trade agreements, as demonstrated for example in the May 10th agreement which was a significant improvement in protecting workers and the environment. Just as we have insisted on a level playing field in our trade agreements, where workers' rights and the environment are protected, we must now ensure that there is no room for companies to relocate to countries with weaker climate legislation than ours. As outlined below, we believe we can accomplish this in a manner that is consistent with our current obligations in the World Trade Organization.

Emissions Leakage is a Critical Issue

I address you today to discuss the critical issue of how to control carbon dioxide emissions in the United States in a way that does not result in economic effects that spur an exodus of American energy-intensive industries.

The Sierra Club supports implementing a cap and auction mechanism that reduces carbon dioxide emissions at least 80 percent by 2050 to meet current scientific expectation of the U.S. share of reductions required to avoid the worst predicted effects of global warming. That mechanism can logically be expected to raise costs for energy intensive manufacturing, although we are working to see a cap designed to minimize increases in energy costs. Leakage occurs if those costs cause that manufacturing base to simply move to nations that do not have similar controls on carbon. Currently we believe there is cause for concern that this will be a problem in the energy intensive manufacturing sector. Given the highly global nature of these industries, it is reasonable to guard against such an outcome in both the domestic and international attempts to address climate change. We want to protect against off-shoring both because we want to maintain our domestic manufacturing base and because we want to protect the environment.

We have already seen a steady growth in emissions related to the production of exports. In 2005, 33 percent of China's domestic CO2 emissions were in the production of exports, while in 1987 this number was only 12 percent. Recent research has also shown that Chinese exports are responsible 6 percent of global greenhouse gas emissions in 2005.[1]

Illegal Logging

An example of how unregulated trade has had severe negative impacts on jobs, communities and the environment can be found when examining the pulp and paper industry, and the connections to trade in illegally harvested timber. Deforestation accounts for 1/5 of global greenhouse gas emissions[2], and protecting the world's forest resources is a cornerstone in preventing the most catastrophic impacts of climate change.

Among the causes for deforestation is the fact that there is a large unregulated market place for wood and timber products with few meaningful incentives for stopping illegal logging and the associated trade in these illegally harvested products. In addition to contributing dramatically to global warming, there are also significant effects on workers and the economy. Over the past few years, thousands of jobs have been lost in the U.S. paper industry as paper mills and converting facilities have been shut down. The paper industry conservatively estimates a loss of $1 billion a year as a result of the import of illegally harvested timber and wood products[3]. ILO studies have found that forest workers in countries such as Peru are frequently working under forced labor conditions[4].

The Sierra Club has worked with the United Steelworkers to address the issue of trade in illegally harvested timber and wood products, including through a 2006 petition to the Department of Commerce asking them to investigate whether the sustained non-enforcement of Indonesia's forestry laws constitutes a subsidy. Since that time, Sierra Club and USW have worked together to explore ways to strengthen U.S. trade laws so they can better address illegal logging.

To combat trade in illegally harvested timber as well any other unfair trading practices that occur, we must aggressively pursue policies that ensure a level playing field between domestic and foreign producers. In addition to utilizing our existing trade rules, we also believe that domestic climate legislation must include measures that eliminate any competitive disadvantage U.S. companies might be faced with as a result of the legislation.

Downstream Products

Any mechanism to minimize penalty to domestic manufacturers for compliance with a carbon cap and auction program must address the problem of goods that arrive in the U.S. containing embedded emissions from non-comparable countries. For example, steel made in China that is converted into automobiles in Korea and that enter the U.S. by ship must be accounted for in a correction mechanism. Strong sectoral agreements within a larger global deal are the best way to address this problem.

The Best Protection for American Workers is a Strong Global Climate Deal

A strong global climate deal negotiated through the United Nations Framework Convention on Climate Change (UNFCCC) process is the best solution to competitiveness. A strong global deal that requires action to reduce emissions by all the largest emitters in the energy intensive manufacturing sector should be the goal of U.S. domestic policy, as well as the goal of our international negotiators. To reach this goal we need to bring a balanced proposal to the Copenhagen UNFCCC meeting in December of 2009. This package should include a strong domestic emissions reduction target, and we believe that the science justifies an absolute domestic emissions reduction from of at least 8-14 percent below 1990 levels by 2020.

However, a balanced international proposal will require more than just a domestic cap on greenhouse gas emissions. A full package should include financing for clean energy technology deployment, protections for forests in developing countries, and adaptation to unavoidable climate impacts, including a robust U.S. program of international global warming assistance for developing nations. These additional elements are essential components of the Bali Action Plan, negotiated last year by the Bush Administration through the UN process. In exchange for these elements, developing nations – including China -- have agreed to take measurable, reportable, and verifiable actions to reduce global warming.

There are Several Options Available to Domestic Policy Makers

The only real solution to the problem of emissions leakage is a global agreement by all major emitting countries to reduce emissions from the manufacturing to a negotiated level.

The global deal negotiated in Copenhagen may include all of the building blocks necessary to forge a comprehensive solution to the energy intensive manufacturing sector. In order for a global climate deal to succeed, the U.S. must resume a leadership role in tackling climate change and demonstrate a significant commitment to actually reducing our own emissions. We can accomplish this by passing comprehensive climate legislation which puts a cap on our emissions, incentivizes investments in the green energy economy, and ensures that American companies are not put at a competitive disadvantage vis-à-vis countries that do not take similar action.

There are several options available to policy makers seeking to address the competitiveness problem. The goal of the competitiveness provisions in U.S. climate legislation should be to (1) prevent a decline in output or efficiency by U.S. producers in the face of higher costs, (2) guard against migration of U.S. manufacturing to other parts of the world, 3) ensure that no company reaps a windfall profit, while 4) not losing track of the big picture need to create incentives for other countries to reduce emissions.

It is possible to protect domestic manufacturers with a relatively small percentage of the value of the credits in the system. The main industries that comprise the globally competitive energy intensive sector (iron and steel, aluminum and copper, nonmetal mineral products (cement and glass), paper and pulp, and basic chemicals) account for more than half of all carbon dioxide (CO_2) emissions from the manufacturing sector, though their direct emissions account for less than 6 percent of the U.S. total (WRI 2007). Efforts to solve this problem should be focused on these industries, however Congress should delegate to an agency some responsibility to ensure that any company compensated is in fact subject to global competition.

Allocation of Free Credits

Past climate bills have addressed increased costs to manufacturers by allocating free carbon credits to the globally competitive energy intensive manufacturers based on historical emissions. The concern with this approach is that it does not provide a direct incentive against off-shoring, and instead may increase profits that could be used to offshore later.

Any allowance-based allocation should provide a direct incentive to increase domestic employment and drive investments in the energy efficiency of the company. It should also be designed to avoid windfall profits, since it is possible that those profits could be used to finance off-shoring later. The best way to do this is to tie the rebates to the production-based output of the company and not directly to the pollution.

Output-Based Rebates

Another option being discussed is the use of output-based rebates to compensate energy-intensive manufactures for goods subject to international trade. This option has been suggested by Reps. Jay Inslee and Mike Doyle in H.R. 7146 from the 110th Congress. The Inslee-Doyle framework intends to encourage energy efficiency on the part of manufacturers by setting aside 15 percent of total allowances to compensate them not just according to their cost of compliance with a carbon cap, but also by their emissions intensity and electric efficiency. The successful application of output- based rebates would significantly reduce the possibility that companies can use carbon regulation as an excuse to offshore manufacturing facilities and American jobs.

An output-based rebate could be further improved by allocating only to the average international production per unit of pollution. With this system companies would not be compensated for emissions above the international industry average, giving them a further incentive to improve efficiency. If this approach proves too complicated it may be possible to benchmark to an international standard and then require efficiency improvements to qualify for a full allocation of allowances. In either case, allowance levels should decline over time, transitioning to full auctioning.

An important question is whether it would go too far. While the output-based rebate approach has the advantage of encouraging better carbon performance, it is complex and shares with the conveyance of free allowances the risk of providing windfall profits to manufacturers who have the ability to raise prices in response to the enactment of a carbon control program. We also need to determine if the act of rewarding efficient producers would encourage inefficient producers unable to compete to leave the country rather than upgrade domestic facilities.

Border Taxes and Adjustments

A border tax or adjustment will be needed if industries in other countries fail to take comparable action to address climate change. A correcting mechanism may also be needed to protect against movement of manufacturing to countries that do not have comparable requirements on their domestic manufacturers even if a fair deal is reached with the major emitting countries in Copenhagen. Border adjustment measures only apply to imports to the U.S. and do not fully address competition in the global market where U.S. producers may face greater competition from foreign producers. The only complete solution to this problem is a global deal that addresses the manufacture of globally competitive goods.

We believe that a border tax or adjustment is best set up as a backstop against the failure of a global deal or to address emissions from countries that have not taken similar action by the time we phase out rebates or free credits. While a border mechanism would be part of the legislation from day one, it would phase in as other mechanisms phase out. Negotiating international sectoral agreements is also consistent with WTO rules, which require good faith negotiations before any border tax can be imposed.

Border corrections should not be applied to the least developed countries or to countries whose industries are determined to have taken comparable action (see below).

Border Taxes

A border tax would amount to replacing a cap and trade system with a carbon tax for manufactured goods from internationally competitive energy intensive industries. A national tax would be assessed on the emissions associated with the energy use of certain categories of goods and adjusted over time to achieve reductions comparable to that sought by a carbon cap. Even if the cap is lifted for internationally competitive energy intensive industries the emissions from this sector should be reflected in the reductions required by companies under the cap. The cap should be adjusted to reflect progress toward emissions reductions for taxed industries. The tax would be assessed equally on all emissions-intensive goods consumed domestically, including imports, based on their associated emissions. For imported goods, associated emissions could be calculated based on either firm-specific data or national averages.

There are already examples of how such border taxes have worked to protect the environment, while also being allowable under our WTO obligations. In 1989, as part of our obligations under the Montreal Protocol which was designed to protect the ozone layer by phasing out the production of ozone depleting chemicals (ODCs), Congress adopted production and consumption caps on ODCs, and enacted an excise tax on ODCs at the same time. The tax was imposed on the sale or use by a manufacturer, producer, or importer of any ODC, as well as on any product imported into the U.S. in which an ODC was used as a material in its manufacture or production.

The ODC excise tax was assessed on the same basis for domestic goods sold in the U.S. and for imported goods sold in the U.S. via a border adjustment tax on imports. The taxes were rebated on exports. This consumption-based excise tax drastically curtailed the use of ODCs. In 1990, the year the ODC excise tax was imposed, total CFC consumption dropped to 440 million pounds down from 700 million pounds in 1989[5]. It spurred industry to develop and use alternative chemicals and technologies, effectively protected the ozone layer, and was also compliant with GATT and WTO rules which allow indirect taxes to be adjusted at the border.

Border Adjustment

A border adjustment would use the apparatus of the cap to create the same kind of correction mechanism. Rather than pay a defined tax, it would require companies without comparable carbon restrictions to buy allowances at the border that reflected either firm-specific data on carbon emissions or national averages from the home country. Again, this is a potentially complex route and it does not have a WTO track record to give confidence of compatibility. But it should be possible to design such a system in conformity with WTO principles.

In addition, by virtue of the fact that most of our trade in energy-intensive goods is currently conducted with carbon-comparable nations, a border adjustment might create an increase in the proportion of such trade that is done with Canada and the European Union.

Comparable Action and Sectoral Agreements

We believe the issue of what constitutes comparable action is best addressed on a sectoral basis for developing countries and an economy wide basis for developed countries for the next commitment period of the international climate deal (2013 to 2017/ 2020). This approach is fair in that it allows developing countries to address the parts of their economy that they are most able to address, while still requiring more from the developed world for now.

The short term goal of any competitiveness approach should be to promote international sectoral agreements. Under a sectoral agreement GHG standards would be negotiated within energy- intensive globally-traded sectors. For example, major steel-producing countries could agree on standards limiting GHGs per ton of steel, which could be differentiated initially according to national circumstances and converge over time. These agreements could form the basis for a cap on emissions from the sector in the near term and ultimately would inform economy-wide caps for developing countries. If international sectoral agreements initially require less of developing countries than domestic manufacturers, compensation for energy-intensive industries could be maintained at some level and phased out as the requirements for other countries rise to those borne by the United States.

Conclusions

A combination of rebates or allocations and border taxes or adjustments could effectively level the carbon playing field, but they will not create incentives for developing countries to reduce their domestic emissions or to cooperate in the negotiations. In order to successfully address concerns about the competitiveness of U.S. industry we must also negotiate a fair international climate treaty this year in Copenhagen. Both of these goals are made more achievable if we deliver a balanced and comprehensive climate bill.

Thank you for this opportunity to address the Subcommittee. We stand ready to address any questions you might have.

End Notes

[1] Weber, C.L., etal., The Contribution of Chinese Exports to Climate Change, 2008

[2] Nicholas Stern, The London School of Economics and Political Science, *Key Elements of a Global Deal on Climate Change*

[3] America's Free Trade for Illegal Timber, June 2006, Environmental Investigation Agency, http://www. eia-international.org/files/reports118-1.pdf

[4] Alvaro Bedoya Silva-Santisteban and Eduardo Bedoya Garland, International Labour Organization, *Forced Labor Wood Extraction in the Peruvian Amazo*n, 2005

[5] Elizabeth Cook, ed., Ozone Protection in the United States: Elements of Success, World Resources Institute (Nov. 1996)

In: Carbon Leakage: Trade Issues in Emission Reduction
Editor: Victoria T. Greenwald
ISBN: 978-1-61728-919-4

Chapter 7

STATEMENT OF JOOST PAUWELYN, PROFESSOR OF INTERNATIONAL LAW, CO-DIRECTOR OF THE CENTRE FOR TRADE AND ECONOMIC INTEGRATION, GRADUATE INSTITUTE OF INTERNATIONAL AND DEVELOPMENT STUDIES, GENEVA (SWITZERLAND) AND SENIOR ADVISOR, KING & SPALDING LLP*, BEFORE THE SUBCOMMITTEE ON TRADE OF THE HOUSE COMMITTEE ON WAYS AND MEANS

Joost Pauwelyn

Chairman Levin, Ranking Member Brady, and distinguished members of the Ways and Means Subcommittee on Trade, I am pleased to participate in today's hearing.

I understand that today's hearing is focused on how U.S. Climate Change legislation can limit both carbon leakage and job leakage, that is, maintain U.S. competitiveness.

What I hope to add to today's discussion is how these efforts could be undertaken in line with U.S. obligations under the World Trade Organization (WTO). By way of background, I have worked for the WTO, advising dispute panels and the WTO Appellate Body, from 1996 to 2002. After that, I became a tenured, full professor at Duke Law School here in the United States. In 2007, I returned to Switzerland to take up a professorship at the Graduate Institute of International and Development Studies in Geneva. At the same time, I have been advising the Washington office of the law firm of King & Spalding on matters of trade, investment and climate change, helping, among other clients, energy-intensive industries based in the United States to prepare for tomorrow's low-carbon economy. In early 2007, at the request of Duke's Nicholas Institute for Environmental Policy Solutions, I published a paper on *U.S. Federal Climate Policy and Competitiveness Concerns: The Limits and Options of International*

* The views expressed in this testimony are my own and not necessarily those of the Graduate Institute, King & Spalding or any of its clients.

Trade Law.[1] This paper was referred to extensively in, for example, last year's work of the House Committee on Energy and Commerce.[2] In 2008, I also edited a special issue of the Carbon & limate Law Review on the topic of *Climate Change in a Global Economy*.[3]

SUMMARY

My core message to you today is this. First, WTO rules are flexible enough to deal with both carbon and job leakage. People should stop using the WTO as an excuse not to tackle climate change. Second, even though the WTO should not stand in the way, fighting climate change will have costs, and addressing carbon and job leakage in a cost- effective and administratively feasible way will not be easy. The devil will be in the detail. Solutions to address carbon and job leakage must be carefully calibrated, and may vary, based on industry and country-specific data. In this process of finding appropriate remedies, WTO rules do have a useful, *positive* role to play, that is: To avoid wasteful protectionism or discrimination that serves neither the environment nor American jobs.

In sum, the WTO should not stop your efforts, but merely guide them as one relevant factor amongst many other, arguably more important, criteria such as cost- effectiveness, administrative feasibility and risk of circumvention.

CARBON AND JOB LEAKAGE ARE REAL BUT SHOULD NOT BE EXAGGERATED

Although carbon and job leakage are often exaggerated, it is clear that they are a problem. If today the United States puts a limit on carbon emissions and countries like China or India do nothing, then evidence indicates that the production of some carbon- intensive goods, together with the jobs and emissions associated with them, will shift or leak from the United States to these other, unregulated countries. This leakage rate varies widely between sectors. For low-carbon products, the impact will be negligible. For some carbon- or energy-intensive industries the impact will be real.

One recent study finds that a price of $21 per ton of CO2 applied in Japan and the E.U. - 15 (but not in other countries) would lead to a leakage rate of 55 per cent in the iron and steel sector. Another study finds that a price of $20 per ton of CO2 in a cap-andtrade system applied in the E.U.-27 offers leakage rates between 0.5 and 25 per cent in the iron and steel sector, and between 40 and 70 per cent in the cement sector. The exact numbers depend, among other things, on how allowances are distributed and the capacity for producers to pass-through higher production costs onto consumers. More pass- through capacity means less leakage: If I can pass-on the cost of carbon to consumers and consumers continue to buy, my company will not suffer. Pass-through capacity and thereby leakage is, in turn, determined by the intensity of foreign competition, market concentration and tightness, and the availability of substitutable products. This means that widely traded, fungible products such as cement, aluminum and steel will suffer more leakage than products locally produced and consumed.[4]

To give but one example in the European context as applied to the cement industry, at a price of 25 euros per ton of CO2, the cost of direct CO2 emissions for European cement producers (assuming all allowances were auctioned) would be 20 euros per ton of cement. This represents approximately 40 per cent of the total cost to produce a ton of cement. If this additional 40 per cent cannot be passed-through to consumers, many European cement producers will simply have to close shop.[5]

So, to the extent that carbon and job leakage is, indeed, a problem, what can we do about it?

The First-Best Solution Is an International Agreement to Limit Emissions across Countries

The first-best solution is, no doubt, to conclude international agreements with countries such as China, India and Brazil that mandate those countries to also cut emissions and to price-in the cost of CO2 into their production processes.

When it comes to such international agreements, we must be realistic and fair. Given historical emissions and the head-room needed for developing countries to build up their economies, emission reductions should not, and will not be, the same for each and every country. The United States has ratified and is a party to the UN Framework Convention on Climate Change (UNFCC) (albeit not, of course, to the Kyoto Protocol) and in this convention all countries, including the United States, agreed to the principle of "common but differentiated responsibilities and respective capabilities".[6] This principle was, moreover, reconfirmed in the July 2008 G8 Hokkaido Toyako summit declaration.

Thus, if, for example, China were to agree to carry its share of the burden and to cut its emissions to a level that is fair and justifiable, Chinese producers would already bear the cost of carbon under Chinese legislation. In that case, and even though the cost would not be the exact same, both U.S. and Chinese steel and cement producers would be on something of a level playing field, carbon and job leakage would be limited and there would be no need to impose "carbon equalizing" tariffs or other requirements on Chinese imports.

Achieving "carbon equivalence" through such an international agreement would have the added benefit of avoiding WTO litigation or any other trade wars. As Pascal Lamy, Director-General of the WTO put it:

> I am of the firm conviction that the relationship between international trade – and indeed the WTO – and climate change would be best defined by a consensual international accord on climate change that successfully embraces all major polluters. In other words, until a truly global consensus emerges on how best to tackle the issue of climate change, WTO Members will continue to hold different views on what the multilateral trading system can and must do on this subject. Only a consensual international accord can take them forward.[7]

We must, however, also be realistic in another way. Notwithstanding the principle of "common but differentiated responsibility", we cannot exclude as a possibility that in the Copenhagen climate change negotiations at the end of this year, some developing countries will simply refuse to carry their share of the burden. If this were to happen, what can countries like the European Community, the United States or Japan do, assuming that these countries do move forward and limit their carbon emissions?

THE WORLD OF SECOND-BESTS: "CARBON EQUIVALENCE" THROUGH LOWER COSTS ON ENERGY-INTENSIVE U.S. INDUSTRIES, OR HIGHER COSTS ON ENERGY-INTENSIVE IMPORTS

Within the second-best world where no international agreement can be reached, two main options are available (options that are, moreover, not mutually exclusive):

- First, the United States could *soften* the impact of climate change legislation on those *U.S. industries* particularly exposed to carbon and job leakage (think of cement, steel, aluminum, etc.). The most obvious solution to this end is for the government to allocate free allowances to these industries (be it based on past emissions or so-called grandfathering, or by using some benchmark of carbon-intensity).
- Second, the United States could *harden* its position against *imports,* especially imports of carbon-intensive products, and impose a carbon cost on those imports to level the economic playing field and thereby avoid carbon leakage.

> The most obvious solution to this end, within a U.S. cap-and-trade scheme, is to oblige importers to buy emission allowances. Another option could be to enact an energy-performance or energy-intensity standard for certain products (say, a ton of steel cannot have a carbon-footprint of more than x tons of CO_2) and to impose that standard on both domestic steel and imported steel. A variation thereof is to set a carbon-intensity benchmark with products producing less than the benchmark obtaining carbon credits and products above the benchmark having to buy credits to make-up for their heavy carbon- footprint. Such "tradable performance standards" (TPS) is what the Coalition for Sustainable Cement Manufacturing & Environment has proposed in the context of state-wide California climate change efforts.[8]

In sum, to achieve "carbon equivalence" across countries and thereby avoid or at least limit carbon and job leakage, three options are available: (i) international agreement, (ii) lower the internal, U.S. cost of carbon, or (iii) impose a carbon cost also on imports (be it through import allowances or a performance standard applied to both domestic and imported products). A combination of these three options seems likely, whereby some countries would be recognized as making "equivalent" efforts as compared to those made by the United States (for example, E.C. imports would not have to buy allowances); some U.S. industries get (more) free allowances (and/or for a longer time); and certain imports from other countries, who refuse to (sufficiently) cut their emissions, have to buy a certain (variable) number of allowances corresponding to their climate change efforts and level of development.

What now would the WTO say about any of this? As noted earlier, the WTO would prefer the first option of international agreement and this option is unlikely to raise WTO questions. Cutting costs for certain U.S. industries, however (Option 2), may lead to a WTO challenge based on the argument that such carve-outs or free allowances for, say, U.S. cement or steel amount to an unfair or trade-distorting subsidy in violation of the WTO subsidies agreement. Imposing a carbon cost, tariff or allowance requirement on imports, on the other hand (Option 3), risks a WTO challenge based on the argument that such added cost on imports discriminates imports as against domestic products (in violation of the national treatment principle) or discriminates against some countries as against others (in violation of the most-favored-nation clause).

Carve-Outs or Free Allowances For Carbon-Intensive U.S. Industries Can Be Modeled in Compliance with WTO Subsidy Rules

The WTO subsidies agreement only prohibits two, very specific types of subsidies: (i) subsidies that are contingent on *exporting* the final product, and (ii) subsidies that are contingent on using *domestic inputs.*[9] Production subsidies, or free allowances that are handed out even if products are not exported or even if foreign inputs are used in the production process, are not prohibited under the WTO treaty. Such production subsidies may only be challenged at the WTO if they can be shown to be particularly trade-distorting, that is, cause "adverse effects" to other WTO Members such as "serious prejudice" to, for example, China or Europe, by impeding or displacing Chinese or European steel or cement on the U.S. or third country markets or "significantly undercutting" the market price of steel or cement.[10]

Would carve-outs or free allowances for energy-intensive U.S. industries constitute such so-called "actionable" or "trade-distorting" subsidies? On this question, four lines of defense are available.

Firstly, it is not even clear that free allowances or lower emission cuts imposed on energy-intensive industries are "subsidies" in the first place. For there to be a subsidy, there must be a "financial contribution" by the U.S. government or income or price support.[11] The argument can be made that although these industries would carry less of a burden than others, the legislation still imposes costs on them (e.g. the opportunity cost of using rather than selling a free allowance[12]) instead of offering them a "financial contribution". In addition, for there to be a "subsidy" in WTO terms, the industries must be given a "benefit" as compared to what normal market conditions would offer them.[13] Here, the argument can be made that the market point of comparison is not other U.S. industries that are hit harder, but rather similarly situated, carbon-intensive industries that are equally treated or exempted. Thus, even if carbon-intensive U.S. industries could be seen as receiving a "financial contribution" in the form of, for example, free allowances, it arguably does not offer them a "benefit" as compared to normal market situations.

Secondly, for a subsidy to be "actionable" under WTO rules, it must be granted to a "specific" enterprise or industry or group of enterprises or industries.[14] If free allowances or carve-outs are granted not just to one company or industry (say, cement), but to all industries based on the objective criterion of, for example, a high risk of carbon leakage, and this eligibility criterion is "automatic", "clearly spelled out" and "strictly adhered to", the argument can be made that the subsidy is not "specific".[15] And if a subsidy is not "specific", it is not "actionable".

Thirdly, and probably most importantly, other WTO Members would have a hard time demonstrating that free allowances or carve-outs cause them "serious prejudice". On the contrary, as noted earlier, carbon-intensive industries in the United States would carry some extra burden and it would be hard to show how this extra burden harms the bottom line of, for example, Chinese companies. In addition, when it comes to countries that do cut emissions, many of them also hand out free allowances or have special provisions for energy-intensive sectors.[16] In this context, to show that similar carve-outs in U.S. climate change legislation cause "serious prejudice" to E.C. exporters may be difficult.[17]

Fourthly, and finally, although the WTO subsidies agreement itself, as of 2000, no longer offers a safe-haven to certain "green subsidies", the standard *General Agreement on Tariffs and Trade* (GATT) does include an environmental exception (GATT Article XX). Hence, even if none of the above arguments were accepted (i.e., there is a "subsidy", it is "specific", and it does cause "serious prejudice"), a last line of defense for the United States could be to argue that free allowances or other carve-outs are necessary to protect the environment in line with GATT exceptions.[18] At first blush one could claim that carve-outs or free allowances will harm the environment by allowing heavy polluters to continue to pollute. Yet, if free allowances can be sold in the market and are limited to historical emissions or calculated based on a benchmark of normal or average carbon emissions, the affected industries will have an incentive to cut emissions. As a result, even free allowances can contribute to lower carbon emissions and thereby protect the environment. It could also be argued that without such carve-outs or free allowances, the United States would not have adopted any climate change legislation at all (which would surely be much worse for the environment). In addition, carve-outs and free allowances are instruments to limit carbon leakage. Without such instruments (i.e. with full carbon costs on U.S. cement or steel) overall, world-wide emissions could actually have increased by means of U.S. production shifting overseas and producing cement or steel in, for example, China with higher levels of carbon than those emitted previously in the United States, not only emitted during production in China (both directly in the production process and indirectly, for example, through unregulated electrical generation), but also with the additional negative impact of emissions associated with transporting these goods long distances to the United States.

In sum, if carefully calibrated along the lines suggested above, carve-outs or free allowances for carbon-intensive U.S. industries can be modeled in compliance with WTO subsidy rules. More difficult, and arguably more important, than achieving compliance with WTO rules will be to design a system of free allowances that provides the right incentives to energy-intensive industries (based on solid empirical, sector-specific studies and predictions; not the strength of their lobbying efforts), contributes to stopping climate change and limits carbon and job leakage, all of this without offering loopholes to avoid making any contribution at all, or making the system so complicated that commitments are ultimately circumvented or simply not contributing to lowering carbon emissions.

Carbon Equalization Measures at the Border on Imports Can Be Modeled in Compliance with WTO Non-Discrimination Rules

WTO rules prohibit two types of discrimination on the ground that efficient trade and buying decisions should be based on price, quality and other market factors, not national origin: first, WTO rules prohibit discrimination of imports as compared to like *domestic* products (national treatment under GATT Article III); second, WTO rules prohibit discrimination of imports from one country as compared to like imports from *another WTO Member* (most-favored-nation under GATT Article I).

In essence, this means that if a U.S. "carbon equalization measure" at the border – be it a border duty, allowance requirement or performance standard[19] -- does not impose a heavier burden on, for example, imported cement as opposed to U.S. cement, nor a heavier burden on, for example, Chinese cement as opposed to Indian or European cement, the "carbon equalization measures" will pass WTO muster.

What is more, even if some discrimination were found (with, for example, Chinese cement having to buy import allowances, but not European cement), or any other WTO rule were found to be violated (such as the U.S. obligation not to impose import duties above a certain ceiling), the WTO explicitly provides for environmental exceptions. These environmental exceptions permit some discrimination and other types of tariff or import restrictions for as long as they are "related to" the conservation of exhaustible natural resources (including clean air and the earth's atmosphere) or are "necessary" to protect "human, animal or plant life or health".

Exporting countries are likely to argue that the United States ought not to interfere with their sovereign decision of how, for example, China wants to protect its environment or to regulate production processes and carbon emissions on Chinese territory. Yet, since carbon emissions do not respect state borders and carbon emitted in China can affect the U.S. environment as much as carbon emitted in the United States, this territorial argument based on sovereignty and the principle of non-intervention is not likely to carry much weight. Climate change is a global problem that affects all countries. Solving this problem cannot be done by one country limiting simply its own emissions. If other countries do not want to cooperate by means of an international agreement to cut emissions (Option 1 above), and the risk of climate change is real, countries who do want to fight it have a right under WTO rules to also condition *imports* based on their carbon footprint and to do so on a unilateral basis, without the agreement of exporting countries. The WTO Appellate Body, the highest judicial organ of the WTO, said as much in the *US – Shrimp/Turtle* dispute (where the WTO eventually accepted a unilateral U.S. ban on shrimp imports on the ground that foreign shrimp were caught in, for example, India or Malaysia, in ways that do not protect endangered sea turtle at the same level as prescribed for U.S. fisherman):

> It appears to us ... that conditioning access to a Member's domestic market on whether exporting Members comply with, or adopt, a policy or policies unilaterally prescribed by the importing Member may, to some degree, be a common aspect of measures falling within the scope of one or another of the [GATT] exceptions ... [these exceptions] comprise measures that are recognized as *exceptions to substantive obligations* established in the GATT 1994, because the domestic policies embodied in such measures have been recognized as important and legitimate in character.

That said, to meet the GATT environmental exception, U.S. carbon equalization measures cannot be applied in a manner that would constitute "arbitrary or unjustifiable discrimination *between countries where the same conditions prevail"*, nor constitute a "disguised restriction on international trade". This language does, however, allow the United States to make distinctions between imports from different countries (say, Burkina Faso as opposed to China or Europe as opposed to India) for as long as "different conditions" prevail in those countries. Most climate change bills before the 110th Congress, for example, stipulated that imports from least-developed countries would be excused as would imports from countries that take "comparable action" as compared to U.S. action to limit emissions. Since these countries are not in the "same situation", i.e., different "conditions prevail", differentiating between their imports would be acceptable for as long as the differential treatment is based on objective factors related to these countries' "situation", and not to the national origin of the products concerned.

To comply with the above guidelines of non-discrimination and/or the WTO's environmental exception, the devil will be in the details. Here are some elements that will play a role:

- Efforts must be made to ensure that imports pay the same price per ton of carbon as domestic, U.S. produced goods. Where allowances are openly traded, and a separate pool exists for import allowances, a system must be developed so that the price of import allowances carefully tracks the price of domestic allowances.
- Besides pricing allowances in a non-discriminatory way, the calculation of the number of allowances to be submitted for each imported product must also be objective and non-discriminatory. For example, where U.S. products benefited from free allowances so must imported products. In addition, WTO rules prefer a carbon assessment on a *product-specific* basis by, for example, allowing an importer to demonstrate the actual carbon-footprint of a specific batch of imports. In the absence of such product-specific information, the United States could use an *industry-wide* historical baseline for each country, compare that baseline to most recent emissions, and divide the excess by the number of units produced in the country in question. The disadvantage of such industry-wide calculation, as compared to a product-specific calculation, is that it may not offer an incentive to, for example, an individual Chinese steel producer to cut emissions since the number of allowances he will have to buy at the U.S. border will anyhow be determined by overall emissions in the Chinese steel sector as a whole. The advantage of an industry-wide calculation, on the other hand, is that it does put more pressure on the Chinese government to enact regulations that mandate emission cuts on an industry or even country-wide basis since only if overall emissions go down (or China takes climate change action "comparable" to that of the United States) will Chinese exports have to pay for less, or no, emission allowances.
- To underscore the environmental objectives of carbon equalization measures at the border, and to avoid discrimination between products or countries, it is crucial that any differential treatment between products (say, computers and cement; raw materials and finished products) or between countries (say, countries with high, low and no climate change action in place, or different levels of development or carbon emissions) are objectively justified with reference to environmental criteria including the risk of carbon leakage. Arguments or indications of economic competitiveness concerns or leveling of the economic playing field between, say, U.S. and Chinese steel, will not carry much weight in the WTO; on the contrary, they would most likely be used in support of a finding that U.S. legislation is protectionist or discriminatory and, therefore, violates the WTO treaty.
- Another consideration when deciding on whether carbon equalization measures on imports are protectionist of U.S. industries (generally not permitted under the WTO) or protectionist of the global environment (generally permitted under the WTO) is what the U.S. authorities would do with the moneys collected when selling or auctioning carbon allowances for imports. If this money goes into environmental

projects, preferably global ones, and importers are, moreover, able to present foreign, equivalent allowances or credits (receipts for which would not even go to the U.S. treasury), this would be evidence of a genuine environmental objective. If, in contrast, the receipts would be refunded to specific U.S. import-competing industries (to be distinguished from refunds to U.S. taxpayers in general), this could present evidence of a protectionist or discriminatory design.

In sum, if carefully calibrated along the lines suggested above, carbon equalization measures at the border, imposed on certain imports, can be modeled in compliance with WTO non-discrimination rules and/or the WTO's environmental exception. It must be clear, however, that such climate change measures on imports can be levied only to impose a carbon cost on imports equivalent to the carbon cost imposed on like domestic products. Such measures cannot be used or abused as a tool to bring overall import prices up to the level of U.S. prices. To the extent imports of, for example, steel or cement from China are undercutting U.S. producers for reasons unrelated to U.S. carbon costs, carbon equalization measures cannot remedy that situation. To do so, the WTO does provide, however, for other tools, namely anti-dumping duties to offset the price-undercutting effect of dumped imports, countervailing duties to offset the injury caused to U.S. industries by foreign subsidies, or safeguard measures to limit all imports because of a sudden surge in imports that causes serious injury to U.S. industries.

To illustrate this point with an example, if the price of U.S. cement is $10 and because of U.S. climate change legislation it rises to $14, and the price of Chinese cement (not subject to emission cuts) is $8, then carbon equalization, assuming both emit the same amount of carbon, implies, at a maximum, an extra charge on Chinese cement of $4 so as to offset the carbon cost on U.S. producers, not an extra charge of $6 so as to equalize overall prices.

Finally, additional charges or allowance requirements on imports do have three drawbacks that free allowances for U.S. industries may not have. First, additional costs on, for example, imported steel from China or Korea will make producing cars in the United States more expensive. Trade barriers on energy-intensive inputs make production in the United States more expensive. Second, measures restricting imports based on their carbon-footprint may be seen by some countries as an incentive to ratify a post-Kyoto deal and to cut their emissions at source. Yet, other countries may see it in less benign terms and interpret it as something that will make international agreement more difficult. The latter may not only provoke blockage in climate change negotiations and WTO complaints. It may also trigger similar import restrictions in other countries, including on U.S. exports into countries that do not regard U.S. action against climate change as "comparable" to theirs. That said, the risk of an escalating trade war should not be overstated. If, for example, China wants to retaliate against U.S. import restrictions on the ground that those restrictions violate WTO rules, China must first file and win a WTO complaint and subsequently receive WTO authorization to retaliate.[20] Whenever a country were to retaliate against the United States on the ground of an alleged U.S. violation of WTO rules without such WTO authorization, the United States can challenge such retaliation as inconsistent with the WTO treaty.

A third drawback of carbon equalization measures on imports, as opposed to lowering carbon costs on U.S. energy-intensive industries, is that such border measures rely on accurately detecting and influencing activity abroad, outside the jurisdictional control of U.S. authorities, and depend on information provided by foreign countries and foreign producers.

Accurately gauging carbon emissions within the United States is difficult enough; calculating the carbon-footprint of foreign operations is even more difficult (albeit not impossible, witness standard U.S. anti-dumping investigations where U.S. authorities similarly rely on information provided by foreign entities). In addition, where U.S. legislation imposes an allowance requirement on some primary products (such as steel) but not on downstream products (such as cars), the extra duty could be avoided by relocating car plants away from the United States, a move that would only exacerbate job leakage.[21] Finally, where U.S. border measures make fine-line distinctions between the number of allowances to be bought by imports depending on their national origin (say, Chinese cement must buy 10 allowances per ton; cement from Korea only 8), producers may find ways to circumvent the higher burden and get their product into the United States not as "made in China" but as "made in Korea". This would neither protect the environment nor protect American jobs.

In closing one possible advantage of import equalization measures over carve- outs or free allowances for certain U.S. industries must be pointed out. Pricing carbon as it is emitted both by, for example, U.S. steel producers and imports of steel from China, does put a cost on carbon even in respect of energy-intensive products. In contrast, carving-out certain U.S. industries or giving them free allowances lowers the cost of carbon for those industries and does not impose any cost on imports. If effective, overall emission cuts under an import equalization scheme could, therefore, be expected to be greater than under a system of free allowances for certain U.S. industries (unless, of course, the absence of cuts in U.S. energy-intensive industries is more than made up with bigger cuts elsewhere). In this respect, a carbon intensity standard that all suppliers, domestic and foreign, must meet (with flexibility to buy/receive allowances for the difference from the standard) might be an attractive alternative.

Even If the United States Were to Violate WTO Rules, the WTO Dispute Settlement System Offers Broad Flexibility

From the analysis above, it should be clear that even if I am convinced that a WTO consistent way of addressing climate change and carbon and job leakage can be designed, WTO law and jurisprudence is still in its infancy and no one can tell with full certainty that something will, indeed, pass WTO muster. Especially if the United States were to impose carbon equalization measures at the border, on imports, a WTO challenge would be very likely. The United States would have very good arguments to defend such measures. Yet, even if the measures were not well designed or otherwise found to be inconsistent with WTO rules, the legal consequences are rather limited.

First, a WTO proceeding is likely to take two years, if not more, and even if the United States were to lose, it would not have to pay any compensation for past harm. WTO remedies are purely prospective.[22] Second, the United States would get a second chance to amend the details of its legislation to avoid, for example, discrimination. It would be given a "reasonable period of time" to do so, and would not have to pay or compensate for anything during this interval.[23] Third, even if carve-outs or free allowances for energy-intensive U.S. industries were found to violate WTO subsidy rules, the United States would not be obliged to "withdraw" those subsidies, but only to stop or "remove" their "adverse effects".[24] Fourth, whatever WTO violation could be found, the United States can, in practice, decide to maintain its legislation and instead pay trade compensation or accept similar trade restrictions imposed by other, complaining WTO members.[25] In the long-standing *Hormone-treated beef* dispute, for example, the WTO repeatedly condemned the E.C. for unjustifiably banning hormone-treated U.S. beef.

Yet, the E.C. got away with this "illegal" ban, first, by suffering U.S. retaliatory import restrictions for years, and second, by offering the United States more market access elsewhere in a move that recently ended the dispute (the United States now has a bigger quota to export hormone-free beef into the EC). The United States did something similar after a WTO condemnation of a U.S. ban on online gambling services from Antigua: Instead of lifting the U.S. ban on online gambling, the United States negotiated to keep the ban in place in return for opening its market elsewhere. This is not to say that violating the WTO is costless (there are important reputational costs) or that compensation packages come for free. Only to point out that the WTO does not have the power to force its members to effectively change their legislation. In this sense, the WTO is, especially for large countries like the United States, a welcome check or control mechanism at the edges (to avoid, for example, harmful discrimination or protectionism) without, however, fundamentally affecting U.S. scope of legislative manoeuvring especially when it comes to such a fundamental interest as fighting climate change.

CONCLUSION

WTO consistent ways are available to fight climate change and to address both carbon and job leakage. If the WTO, in the famous *US – Shrimp/Turtle* dispute, gave the go ahead for the United States to "equalize", in respect of imports, regulations on U.S. shrimp fishing aimed at protecting endangered sea turtle, then surely there must be a WTO-consistent way for the United States to "equalize" the cost of carbon imposed on U.S. energy-intensive industries aimed at saving the planet. That is not to say that these "carbon equalization measures" are easy to design and effectively implement. There is no doubt that reaching an international agreement where all major carbon emitters agree to cut their emissions would be the first-best solution. In addition, or as an alternative, U.S. climate change legislation could impose a lower carbon cost on certain U.S. energy- intensive industries (through, for example, free allowances). As a measure of last resort, U.S. climate change legislation could also impose carbon requirements on certain imports. This should only be a measure of last resort given the additional problems of higher input costs, detecting and influencing activity abroad and possible negative reactions from other countries. Ideally, the threat of such import measures should suffice to guide reluctant countries into a post-Kyoto agreement and back to the first-best solution. If that fails, import measures, as difficult as they may be to design and implement in practical terms, must not be excluded, certainly not on WTO legal grounds, both to protect the environment and to protect American jobs.

End Notes

1 April 2007, Working Paper 07/02, Nicholas Institute for Environmental Policy Solutions, Duke University, available at http://www.nicholas.duke.edu/institute/internationaltradelaw.pdf.

2 Climate Change Legislation Design White Paper, Competitiveness Concerns/Engaging Developing Countries, January 2008, available at http://energycommerce.house.gov/Climate_Change/White_Paper.Competitiveness.013108.pdf.

3 1 Carbon & Climate Law Review 2008 (CCLR), Lexxion, Berlin.

4 Julia Reinaud, Issues behind competitiveness and carbon leakage, Focus on heavy industry, IEA Information Paper, IEA/OECD, Paris, 2008, at 4.

5 Julia Reinaud, Industrial Competitiveness under the European Union Emissions Trading Scheme, IEA Information Paper, IEA/OECD, Paris, 2005.

6 Article 3.1 of the UNFCC.

7 Pascal Lamy, Preface, 1 Carbon & Climate Law Review (2008), 1.

8 In the interest of full disclosure, King & Spalding LLP advised the Coalition for Sustainable Cement and Manufacturing & Environment in this context.

9 Article 3 of the Subsidies Agreement.

10 Articles 5 and 6 of the Subsidies Agreement.

11 Article 1.1(a) of the Subsidies Agreement.

12 Moreover, in many cases free allowances operate within a maximum cap which does not allow for growth, thus giving unregulated imports an advantage. In addition, even with 100 per cent free allowances, cement companies, for example, would still be facing higher indirect costs from climate change legislation like higher power rates and higher raw material costs.

13 Article 1.1(b) of the Subsidies Agreement.

14 Article 2 of the Subsidies Agreement.

15 In support, Article 2.1(b) of the Subsidies Agreement: "Where the granting authority, or the legislation pursuant to which the granting authority operates, establishes objective criteria or conditions governing the eligibility for, and the amount of, a subsidy, specificity shall not exist, provided that the eligibility is automatic and that such criteria and conditions are strictly adhered to. The criteria or conditions must be clearly spelled out in law, regulation, or other official document, so as to be capable of verification".

16 The current E.U. emissions trading scheme permitted for free allocation of up to 95 per cent of allowances for the 2005-2007 period and 90 per cent for the 2008-20 12 period. Switzerland's scheme, which started in 2008, opted for 100 per cent free allocation.

17 Looking beyond 2012, the European Commission has proposed to make auctioning of allowances the principal mode of allocation for, for example, the power sector. Other sectors would be divided in two, distinct categories: (i) those identified as not "exposed to the risk of carbon leakage" which would continue to receive some free allowances until 2020, and (ii) "[e]nergy-intensive industries which are determined to be exposed to significant risk of carbon leakage" which could receive a higher amount of free allocation or for which an "effective carbon equalization system" could be imposed "with a view to putting EU and nonEU producers on a comparable footing" (paragraph 13, Draft Proposal for a Directive of the European Parliament and of the Council amending Directive 2003/87/EC, Brussels, January 2008).

18 Whether GATT exceptions apply also to rules under the Subsidies Agreement remains an open question and has not yet been tested in WTO jurisprudence.

19 Performance or carbon-intensity standards on a product-specific basis may also trigger U.S. obligations under the WTO's Agreement on Technical Barriers to Trade (TBT). Yet, like GATT rules, these TBT obligations center on non-discrimination and the goal that technical regulations or standards should not be more trade-restrictive than necessary to achieve a legitimate objective, in this case, combating climate change. Note in this respect that a carbon-intensity standard would not ban imports based on their national origin but on their carbon-footprint and that under a so-called "tradable performance standard", all imports would be allowed (those not meeting the standard could simply buy additional credits or allowances to make up for the deficit).

20 Article 23 of the WTO Dispute Settlement Understanding.

21 Note that this risk is, however, not present when it comes to imports of cement: Downstream products using cement – such as buildings -- are not normally traded.

22 Article 19 of the WTO Dispute Settlement Understanding.

23 Article 21.3 of the WTO Dispute Settlement Understanding.

24 Article 7.8 of the Subsidies Agreement.

25 Article 22 of the WTO Dispute Settlement Understanding.

In: Carbon Leakage: Trade Issues in Emission Reduction
Editor: Victoria T. Greenwald
ISBN: 978-1-61728-919-4

Chapter 8

TESTIMONY OF ROBERT E. CLAY, CEO AND CHAIRMAN, BOARD OF DIRECTORS OF PRIDGEON & CLAY, INC., BEFORE THE HOUSE COMMITTEE ON WAYS AND MEANS, SUBCOMMITTEE ON TRADE, HEARING ON "TRADE ASPECTS OF CLIMATE CHANGE LEGISLATION"

Robert E. Clay

I would like to thank Chairman Levin, Ranking Member Brady, and the other members of the subcommittee for inviting me to testify today. My name is Bob Clay, and I am President and CEO of Pridgeon & Clay, a middle market independent supplier of metal parts and assemblies, primarily focused on exhaust systems and catalytic converters for the automotive and light truck industry. I am here because as the costs of manufacturing in America continue to increase, my company and employees become less globally competitive. I am concerned that certain proposals to address the important issue of climate change will increase our costs and reduce our ability to compete internationally in an increasingly difficult economic climate.

My connection to Pridgeon & Clay is more than economic. My father started the company in a converted garage in Grand Rapids, Michigan in 1948 and built the company over the next four decades, before he retired. In June of 2008 my company employed over 700 people in Grand Rapids and over 150 in Franklin, Indiana. Due to the current economic climate in the automotive industry we have laid off a combined total of nearly 400 people. While some will return as our industry recovers, many may not, and it is important that Congress not take actions that would further threaten our remaining jobs.

I believe that addressing environmental concerns is critical to our future but would like to clarify that although this hearing is about the trade aspects of climate change, my company views this as a debate not just about climate change, but really about jobs and our global competitiveness. I am concerned that while the current climate change proposals are well intentioned, they risk jeopardizing the 60 years of hard work that went into building our company, and especially the future of our employees in the U.S.

Pridgeon & Clay, like many other companies in the automotive industry, depends upon our ability to supply our customers internationally – and to compete on an international basis – in order to thrive and support our employees. A significant part of our growth has been because we have followed our customers to other countries, and by doing so, we have created additional jobs in our U.S. facilities. Over ten years ago we opened a facility in Hungary to supply parts to the European operations of our customers. Last year, we formed a joint venture in Mexico, again to follow our customers and serve them in foreign markets where their businesses are growing.

Our international expansion has not been an effort to produce low cost products in other countries to be exported back to the U.S. In fact, last year we exported roughly $30 million worth of parts from the U.S. to foreign markets. However, the reality is that if our U.S. operations in Michigan and Indiana are not globally competitive – especially due to high regulatory costs in the U.S. that we do not experience in other countries – then it will be difficult to continue to grow in the U.S.

Pridgeon & Clay is a highly automated, efficient company, but we are also energy intensive, so even a slight increase in operating costs related to energy costs could cause us to lose a contract with a customer. If our costs increase we are vulnerable to competition from abroad.

The climate change policy debate is occurring in the middle of an economic climate that is already pushing U.S. manufacturers to the breaking point. We currently face an array of tough challenges: tight consumer spending, difficulty gaining access to credit, the collapse of car and light truck sales causing problems for automakers and our other customers, and pro-manufacturing policies being developed by other countries around the globe who are actively seeking to increase their own manufacturing base.

The U.S. simply cannot afford to place additional burdens on companies who are doing everything they can to continue to operate in the U.S. unless our trading partners, and especially the low cost manufacturing countries, are prepared do so as well. Many of our largest customers have moved operations to Mexico. To maintain the business and continue to serve those customers, companies like ours have no choice but to follow our customers. There is a component of our exports that we can continue to competitively manufacture and ship from our U.S. plants. However, I fear that implementation of a cap and trade system, and the related tariff that has been proposed on imported goods from countries who have not implemented a similar cap and trade system, will increase our manufacturing and transportation costs to the point that our remaining export business will be vulnerable.

Our first concern about the creation of a cap and trade system for carbon dioxide is based upon the costs associated with such a program, and the impact that those costs will have on our company's ability to compete internationally. The fact that a cap and trade system will increase costs for consumers of energy is beyond dispute. Even if you disregard the indirect costs associated with such a system, the $650 billion estimate for revenues from the sale of allowances under the system are costs that will be passed on to consumers of energy – especially in energy intensive sectors such as manufacturing.

As I mentioned, Pridgeon & Clay is a fairly energy intensive business. Our primary input is stainless steel, which is an energy intensive product. Our stamping presses have large electric motors and pumps, we have large air compressors, many of our parts are welded, some are heat treated and most require cleaning to remove oils and lubricants.

If the U.S. is not joined by the rest of the world – especially by low cost countries – in a cap and trade system, then additional costs in the U.S. will require manufacturers in the U.S. to choose between increasing the costs of their goods, reducing overhead by taking steps such as laying off employees, or finding alternative locations for their manufacturing operations. All of these options are bad for U.S. consumers, bad for U.S. workers, and bad for the U.S. economy.

It is also important for us to discuss the tariff proposal that has been proposed. This would impose a tariff, or "carbon tax," on imported goods from countries that do not have climate change policies similar to the one under consideration in the U.S. This would seem to make sense, but this type of proposal could actually make things worse for companies such as ours, because it would increase the cost of the raw materials we use to manufacture our products – costs we cannot typically pass along. Our company has found that domestic steel output is, at times, too low to satisfy overall demand for steel by U.S. manufacturers, and as a result we sometimes have to import steel from other countries. Placing a tariff on these inputs, which often amount to 60% of our costs, will compound our problems under a cap and trade system because we will be paying a higher cost for our energy and a higher cost for our imported raw materials. This is a formula that will further drive manufacturing overseas, limit any environmental benefits of a cap and trade system because emissions will be relocated rather than reduced, and threaten the ability for our company to remain economically viable in the US.

Products manufactured in the U.S. must be able to compete against foreign products. If foreign producers are able to avoid the additional costs associated with a cap and trade system, then we will be uncompetitive. This concern has even been echoed by supporters of the cap and trade system, such as the Obama Administration's Secretary of Energy, Dr. Steven Chu, who recently noted that, "The concern about cap-and-trade in today's economic climate...is that a lot of money might flow to developing countries in a way that might not be completely politically sellable."[1] Secretary Chu is speaking about a political issue. I don't care about the political aspects of this issue – but I do care, very deeply, about the jobs of my employees in Michigan and Indiana, and those jobs are very clearly threatened by cap and trade unless it is universally applied throughout the world, especially in low cost countries.

We are committed to our employees and to helping Congress and our country work through the current economic crisis. However, our ability to continue to manufacture products in the U.S. is imperiled by policies that increase the cost of energy, transportation and delivery, and our raw materials. We appreciate the need to address pressing environmental problems, but believe that the cap and trade and tariff options currently being discussed put at risk what remains of the U.S. manufacturing sector. I ask that you keep in mind the millions of manufacturing jobs lost these past several years and the millions more at stake, especially those of us in Michigan Indiana.

Once again, thank you for inviting me to testify and for considering our input on these complex and important matters.

[3] John Broder & Matthew Wald. (2009). Big Science Role Is Seen in Global Warming Cure, N.Y. Times, Feb. 11. Available at: http://www.nytimes.com/2009/02/12/us/politics/12chu.html?_r=2

CHAPTER SOURCES

The following chapters have been previously published:

Chapter 1 – This is an edited reformatted and augmented version of the Congressional Research Service publication, report R40100, December 19, 2008.

Chapter 2 – This is an edited reformatted and augmented version of the Congressional Research Service publication, report R40914, January 28, 2010.

Chapter 3 – This is an edited reformatted and augmented version of a United States Government accountability Office publication, report GAO-09-724R, July 8, 2009.

Chapter 4 – These remarks wered delievered as testimony given on March 24, 2009. John J. McMackin on behalf of The Energy-Intensive Manufacturers' Working Group on Greenhouse Gas Regulation, before the House Committee on Ways and Means Subcommittee on Trade Hearing on Trade Aspects of Climate Change Legislation.

Chapter 5 – These remarks were delivered as testimony given on March 24, 2009. Leo W. Gerard, International President, United Steelworkers, hearing on Trade Aspects of Climate Change Legislation, house Committee on Ways and Means, Subcommittee on Trade.

Chapter 6 – These remarks were delievered as testimony on March 24, 2009. David Hamilton, Director of Global Warming and Energy Programs, Sierra Club, before the Subcommittee on trade of the Committee on ways and Means.

Chapter 7 – These remarks were delievered as testimony on March 24, 2009. Joost Pauwelyn, Professor of International Law, Co-Director of the Centre for Trade and Economic Integration, Graduate Institute of International and Developmental Studies, Geneva (Switzerland) and Senior Advisor, King and Spalding LLP, before the Subcommittee on trade of the House Committee on Ways and Means.

Chapter 8 – These remarks were delievered as testimony on March 24, 2009. Robert E. Clay, CEO and Chairman, Board of Directors of Pridgeon and Clay, Inc., hearing on Trade Aspects of Climate Change Legislation, House Committee on Ways and Means, Subcommittee on Trade.

INDEX

J

K

L

M

Q

R

S

T

U

V

W

Z